The *Otter Spirit*

A Natural History Story

by

Judith K. Berg

Judith K. Berg

Illustrations by Michael Reed

www.otterspirit.org

Published by Ulyssian Publications, an imprint of Pine Orchard, Inc.
Book design and cover by Matt Gravelle.
Visit us on the internet at www.pineorchard.com

First Printing: Spring 2005
Printed in Canada.

ISBN: 1-930580-70-3
EAN: 978-1-930580-70-1

Library of Congress Control Number: 2004117358

Printed with soy ink on recycled paper
100% post-consumer fiber
totally treated without chlorine

About the Author

Emerging from a traumatic period in her childhood, Judith K. Berg vowed to invest her life in making a contribution. She didn't know how that would manifest itself, until a university professor invited Judy to conduct observations in an atypical captive environment as part of a study on African elephant behavior. She knew then that her contribution would be towards the preservation of endangered species of the world. Her graduate research on elephant vocalizations and associated behaviors quantified for the first time that elephants communicate with low-frequency sounds that are known to be below the range of human hearing.

Armed with her graduate degree, Judy became a Research Associate at the San Diego Wild Animal Park. After completing ten years' research on African elephants, she directed her behavioral studies to endangered Japanese serow, Chinese goral, okapi, and barasingha deer. Concurrent with her research, she also participated actively in local and regional nature conservation issues.

When Judy relocated to Colorado, she was determined that she would conduct the final project of her career on an endangered species in the wild. With encouragement from a number of wildlife professionals, Judy chose the state-endangered North American River Otter on the Colorado River in Rocky Mountain National Park and surrounds. The many hours spent in this spectacular

environment connected Judy with Native American spirituality and even closer to the wonders of Mother Nature and her children.

From every research project that Judy conducted emerged publications in the scientific and popular literature, as well as presentations to international professional and secular audiences. Although she fulfilled her life's goal, her personal drive and commitment led her to write *The Otter Spirit.*

Judy, her husband David, and their special companion Rusty are currently enjoying their retirement—and continuing to contribute to the world of natural history—in Eugene, Oregon.

Acknowledgments

Many individuals contributed to both my academic pursuits and professional career. My gratitude goes out to those college professors and wildlife professionals who gave me their guidance, support, encouragement, and direction. Prior to the river otter study, I conducted my behavioral research primarily at the San Diego Wild Animal Park in Escondido, California. This atypical captive environment opened a window to the world of those wildlife species that allowed me to enter. The friendship of many people at the Park was important to me during those years. I thank them and I thank the animals—African elephants, Japanese serow, Chinese goral, okapi, and barasingha deer—that shared some of their secrets with this curious observer. I particularly appreciate the friendship and assistance of James Dolan, Randy Rieches, Lance Aubery, Andy Blue, Steve Friedlund, Gloria Kendall, Sally Lahm, Rich Massena, Jeannie Plumer, Esther Rubin, and Barbara Schwankl.

Upon entering into the natural world of the river otter, I knew not where the path would lead. Mother Nature greeted me with extended arms that I embraced with my being, my heart, and my soul. Thanks to the river otters, to their precious habitat, and to all of Mother Nature's flora and fauna for guiding me through this incredible world. I vow to do all that I can to ensure this world will prevail throughout time.

I feel deep respect for the Native American peoples and for their connection to the natural world. I thank them for sharing their special stories about Mother Nature and her children through their myths and legends.

In my study area, at the headwaters of the Colorado River in Rocky Mountain National Park, I am grateful to the many Park staff, volunteers, and visitors who took the time to document otter sightings and signs for my project. A special thank you to the merchants and residents of Grand Lake, Colorado, who made me feel at home during the years of my otter project: from Dave and his friendly checkers at the grocery store to the employees at the local gas station to the guys who pulled my truck out of an icy ditch during the middle of winter. From the West Unit of Rocky Mountain National Park, I specially thank Jim Capps, District Interpreter, for his empowerment, assistance, and encouragement during my project. I also thank West Unit staff and volunteers—especially Barb King, Harry Canon, Debi Claassen, June Copeland, Darwin Spearing, Willie Wharton, Jock Whitworth, and Diana Wiggam—and Colorado Division of Wildlife biologist Jerry Claassen. Their friendship and moral support never waned, even during some rather discouraging times. I am equally grateful for the friendship of Lorene and the late Cecil Brooks, Arapaho National Forest hosts at Monarch Lake.

I thank the Board of Directors of the River Otter Alliance for allowing me to share some of my otter writings in the *River Otter Journal*.

I am particularly grateful to my friend and colleague, the late Elaine Anderson, for her personal companionship, professional advice on studying river otters, and encouraging words during my project. My all-too-brief acquaintance with Dorothy Wathen, a Medicine Woman of the Blackfoot Tribe, will always be remembered. I know her spirit kept me encouraged throughout my research and was a guiding force in the writing of this book. She will always be a part of me. Thanks to my mother, Evelyn Duncan, who first introduced me to Nature's garden— and to my life-long friend, "Aunt Lydia" Brown, for her devotional encouragement; her words still mean so very much to this quiet, shy individual.

The critique and constructive suggestions for this manuscript from my friends Meredith McPherson (two drafts), Gloria Kendall, and Tracy Johnston were invaluable in lifting it to its current stature. Thanks also to Elizabeth Lyon for her review of an early draft and to Renny Rapier for his review of a later one—and a special thank you to Carolyn Gravelle of Pine Orchard publishing, whose skillfully applied finishing touches preserved the message and enhanced the telling of my story.

I extend a very special thank you to Eugene, Oregon artist Michael Reed, whose creative illustrations bring visual life to the *dramatis personae* of the story and to Matt Gravelle, who artfully blended Michael's otter portrait with my photograph of the Colorado River in autumn into an imaginative and inspiring cover design.

Last, and most important of all, I thank my husband David Berg, to whom I dedicate this book. David, you have loved, encouraged, and supported me throughout my professional career and far beyond. Your advice and editing skills on my scientific papers—and through many drafts of this book—are most appreciated. You ensured that my periods of doubt and discouragement never stood in the way of my need to contribute with my life. My heartfelt thanks to you.

—Judith K. Berg, 2005

To David

My husband,
best friend,
and pillar of strength.

Contents

Preface

This is a natural history story of the life of a river otter and the autobiographical complement of the otter's researcher. Their unique journey through life transcends two worlds: the real and the spiritual. Travel with them through seasons and years as they discover many remarkable mysteries of the natural world, evoking wonder at its beauty and diversity.

The story of *The Otter Spirit* is based on true accounts of observations the researcher recorded firsthand, and on imagined events that very credibly could be observed, based on reports documented in the referenced literature. The dialogue and expression of feelings among the otters and other animals in this book are admitted anthropomorphic interpretations of events told in the story. However, facts articulated in the book are real and substantiated by direct measurement by the researcher or authenticated in the references.

Hope prevails that, through the tale of the Otter Spirit, a genuine appreciation of these marvelous animals and the environment they call home will be conferred to the reader. For some, a greater passion will drive them to research the literature and, perhaps, go beyond, to find their own Otter Spirit—on a river, in a forest, on a mountain yet to be explored—and personally realize the sacred gifts that Nature offers those who take the time to hear the Ancient Voices.

CHAPTER ONE

The Beginning

From Nature's womb, you'll come to me
to teach of water, land, and tree;
there, you'll enter into my soul—
to become One will be our goal . . .

The Beginning

Mother Nature continues to spread her coat of white across the land. Her delicate crystal tears fall like soft petals gently kissing the Earth.

Spring emerges late in the high mountains, but it won't be long before the warmth of Nature's smile will melt her winter garment, baring the flesh and bones of her firmament. Purified and nourished for emerging flora and fauna, timed to perfection from millennia of natural selection, from her womb will sprout a bounty of new life during this most precious time of year.

You see, according to Old One, the Earth was made from Woman.

High on the western slope of the Continental Divide in the Rocky Mountains of Colorado, near the headwaters of those Great Medicine Waters known as the Colorado River, preparations are being made for the birth of one of Nature's children to whom her life-sustaining force of water is most important.

It is the North American River Otter.

The prospective mother otter prepares the inside of her den with soft vegetation to cradle the new life she will soon bring into it. Because river otters do not construct or excavate their own dens, she may use a lodge or bank den constructed by beavers; or a den excavated by a larger animal, such as a coyote, or by a smaller animal, such as a muskrat, that she could then enlarge; or a natural area like a log jam, hollow tree, or a rock formation; or even a man-made structure, such as a boat house or under a boat dock.

The otter isn't lazy, just very resourceful. After all, except for the first few weeks of her newborn's life, she'll probably not use the same den for a prolonged period of time.

* * *

With pen in hand and a curious mind, I was led in my professional career as a wildlife researcher on an exciting journey of discovery. For many years, I conducted non-intrusive behavioral research on a number of exotic, endangered species residing in an atypical captive state. I was fortunate to travel to the native lands of some of these extraordinary animals where I observed them in the wild for brief periods of time.

Now, nearing the end of my career, I yearn to contribute to yet one more species—this time completely in its natural habitat. With a move to Colorado, I choose the state-endangered North American River Otter.

Even though wildlife professionals tell me that the likelihood of frequent otter sightings is small, and my literature review endorses that fact, I fully expect to defeat the odds and perform an in-depth documentation of their natural behavior.

Professional goal aside, what I don't anticipate is the unique journey that lies ahead. Perhaps what occurs at the onset of my venture should give me a clue:

During the beginning of my project, I meet a Native American medicine woman of the Blackfoot Tribe. Although we spend only a short time in each other's physical presence, she bestows me with some incredible gifts.

She reaches deep into my soul with her childlike passion for discovery. Her excitement reigns supreme, whether we encounter an animal sign or the animal itself. We trek through the forest and along the banks of the river. There, she teaches me to listen to Mother Nature's plea for her children.

I promise to heed Nature's call.

While at the river's edge, my Native American friend performs a special ceremony for me. When she returns home, she performs an additional secret ritual, and then sends me a gift: a small medicine bag.

She encourages me to keep it with me all through my project to guide me to otters and protect me on my journey.

Maybe my need to collect data clouds the meaning of those special gifts during the first year of my project. Whatever the case, research reality sets in as I trek the river and its tributaries to find only otter tracks and scat, slides and tail-drags. At least, I know the otters were there.

Then, one day, my luck changes. I discover an otter den!

This particular den is located about 500 feet inland from a section of the Colorado River headwaters, where a small inlet empties into a large pond. This pristine environment is aesthetically pleasing to the human eye. But more importantly, it contains all the necessary elements for comfort and security of a mother and infant otters.

Can it be a natal den?

The den is externally well concealed. I desire to investigate its interior, but feel concern that someone may be tucked inside. I decide to wait and look later in the year.

But I picture what it may look like . . . and my educated assessment is later substantiated.

I find a system of tunnels and shelves, extending back into the Earth from the entrance. This offers the otters security and a venue for eventual exploratory adventures by newborn pups. The den is located in an area of good land cover. Natural camouflage includes vegetation, hollowed logs, and twisted roots—all providing protection for a mother otter during her daily foraging expeditions between the waters and the den.

My discovery occurs at the end of March, the time for otter births in this section of the country. Snow still carpets the land and a recent heavy storm adds yet more challenge to the forward progress of my small stature. Walking through stretches of willows with snowshoes strapped to my feet, I still sink—almost down to bare ground—through the softer snow surrounding these shrubs. Snowshoe leg lifts become physical energy vampires.

"Oh, to be an otter," I mutter.

The river segment flowing through my study site is at the headwaters and not very wide. Typically, many headwater locations still contain a covering of ice at the beginning of spring. This section of the river, however, has fragments of open waters, and with some ambient daytime temperatures above freezing, I hear snow and ice crashing into its flow.

Slides, tracks and scat found near the river led to my discovery. More heavy snow predicted for later today would have covered the otter's signs. Nevertheless, for some peculiar reason on this particular day, I feel compelled to walk that segment of my study site's forty-mile stretch of watershed.

Is a special newborn otter calling out to me?

I take measurements of the signs and their distance from each other and from the waterway. I then collect the scat, a clue to the otter's diet, for analysis. Once completing this work, I sit down and try to sense the adult otter that leaves me this valuable information.

Moving my hands over the signs, strange feelings enter my being.

What is their message?

* * *

At the end of March, our denning female delivers two small furry pups. Although possible for her to birth six young, normally only two or three are born in the wild.

Her blind, toothless, and helpless babies weigh 3 to 6 ounces, and from the tip of their noses to the end of their tails measures all of 8 to 11 inches. Soft, silky, dark-gray to black fur covers their bodies. This coat, along with their cozy home and the warmth of Mother Otter's body, helps to keep them warm.

Following their birth, the young otters remain in their natal den for seven to eight weeks. In this secure environment, Mother Otter nourishes them with her rich milk and bathes them with her wet tongue. Their eyes will begin to open at four to five weeks of age; at that time, they can explore their den and play with each other.

Many adventures await them when emerging from Earth Mother's womb. Both instinctive and learned behavior governs what lies ahead for them. But for now, sleeping, nursing, playing, and growing are their only requirements.

* * *

As the light of day concedes to the dark of night, I return to my own den and retreat into my cradle of down. I imagine the baby otters curled within Earth Mother's womb as they also seek to rest from their day's activities.

Soon, one of these two furry little animals will distinguish itself in a very different way. This special otter and I are about to enter two worlds: the real, with its accompanying life and death struggles; and the spiritual, with its accompanying comfort and solace.

But what will guide us together?

CHAPTER TWO

The Emergence

Mother Nature has opened her coat of white,
much to her flora's and fauna's delight;
since this is now the emergence of Spring,
even her creeks and her rivers will sing . . .

The Emergence

Nearly eight weeks have passed since the birth of the two otters. Melting ice and snow continue to feed the flow of the river, now at a much more rapid rate as most ambient daytime temperatures reach above freezing. Every day, the river yields a new visual and aural experience. Slowly at first, faster now, Mother Nature releases her grip on her winter bounties. Tumbling into her Great Medicine Waters, they add to its life-giving resource as it propels towards its final destination. Patterns of intense energy intersperse with intervals of serene tranquility as Nature's elements conduct their ancient rituals.

Springtime in the Rockies is that special time when Nature treats all our senses to her spectacular emergence of life. Her majestic mountains shed their coats of white, exposing rugged grays mixed with shades of green. Her soil springs forth with a bounty of flora emerging from its winter sleep, soon to explode into shades of yellows and greens dotted with bonnie colors. Pockets of budding emerald aspen leaves restore their contrast with the rich verdant ponderosa pines that patiently awaited their renewal all winter. The waft of her breath carries fresh fragrances of pines and sugar-sweet flowers across the land. Rhythmic rains, whispering winds, flowing waters, and melodic voices of her avian chorus coalesce to fashion a symphony of synergetic harmonies that fill the air.

This is the time of year when young emerge from the wombs of Nature's fauna. Each newborn will play its own special role in her complex web of life. So much awaits their exploration. But for now, their major role is to play in light-hearted games. Their antics are uninhibited and natural.

Ah, wouldn't it be great to join them?

* * *

The baby otters' restlessness sends a message to Mother Otter that the day has come for them to emerge from the security of their den into the unfamiliar world without.

Encouraged by their mother, two furry faces peek out from the entrance of the only world they have ever known. All their senses work to evaluate the excitement of this new adventure in contrast to the comfort of their den.

The little female appears hesitant, but the male emerges a bit farther. The unusual appearance of a small, asymmetric, white streak on his head above his eyes distinguishes him as a special individual. Their mother calls with reassuring sounds, beckoning them to follow her. Eventually, both move beside Mother Otter ready for the adventure.

She cautiously moves across the land with her two little ones trailing close behind her. She stays constantly alerted to the sights, sounds, and smells which surround them. Otters do not have good visual acuity, but they can still detect movements on land at a considerable distance.

Their eyes are much better adapted for underwater vision. Their acute senses of hearing and smell, however, also play a vital role in their communication system—a system that will mature as the young ones' journey continues.

An abrupt increase in the wind brings new and strange impressions to the otters' keen senses. Dried leaves broken from the snow's caress now rise and fall, teasing the little otters as they brush by. The female moves closer to her mother for security. The male bravely bats at the leaves with his paw.

Yes, they must be the enemy.

As the wind howls, whistles, and sings through the tall trembling pines, even the brave male moves closer to his mother. She senses the heightened fear of her little ones as she turns around and leads the close twosome back towards their den.

Once inside, the young ones nurse on Mother Otter's rich milk, then take a much-needed nap. Although the safety of their den provides sanctuary, perhaps secretly, they look forward to their next outing.

* * *

I continue trekking daily within my study site, trying to be as elusive to the otter as the otter is to me. I don't know the exact date of the new family's emergence. I hope only to observe some of their adventures. In the meantime, I must continue my survey looking for otters and their signs; that's my job. However, each day's outing holds many adventures involving a lot of different species.

With the end of May approaching, migrant birds return to the mountains to play out their ancient rituals that lead to the next generation. Males defend their territories while displaying their brightly colored plumage to attract receptive females.

There goes a robin carrying nesting materials. A hummingbird dive-bombs my head.

"This is my hair, not nesting grass," I respond to the bird's actions. "Ah, maybe you tiny jewel of the sky are just defending your territory."

I sit down at the river's edge to reflect on the surrounding environment and splash some medicine waters onto my face. A bluebird quietly enters the scene and perches on a nearby willow. Gazing up into the azure sky and out onto the blue waters, I realize how they mirror the beautiful color of this bird.

On the landscape along the river's edge, I discover some nearby coyote tracks. Closing my eyes and moving my hands over the tracks, I reflect on lessons my Native American friend taught me on always being aware of signs from Mother Nature. I find myself listening to a legend of the Kalapuya Tribe of northwestern Oregon about the origin of our precious waterways:

> *A long time ago, the frog people had all the water. Whosoever wanted water for any purpose must first go to them. At least, that was the case until trickster Coyote used his cunning and intelligence to change the situation.*
>
> *Coyote went to the frog people with a large dentalia shell in exchange for a big drink of water. The gift was accepted, so Coyote began drinking. Actually, during the long period he had his head*

underwater, he was digging out underneath their dam.

Finally, he rose. The dam collapsed, and the water dispersed—making creeks, rivers, and waterfalls.

Thanks to Coyote, not just one species, but all life forms, can use and enjoy this precious resource.

I open my eyes. Although I sit alone, the river is flowing through my world.

Uh-oh, here comes the wind again—the same wind that sent the otters scurrying back to their den.

I rise to my feet as the bright blue sky dims. Sleeting rain starts descending from emerging dark clouds. What prototypical springtime weather for the Rockies. Sometimes it snows, sometimes it rains, sometimes it hails—and sometimes it does all three!

Now in a soggy state, I desire to retreat into an otter den. Actually, that is not realistic; so instead, I seek temporary shelter cradled within the upper roots of a large pine tree. Mother Nature's arms hold me close to her body. However, one of her children doesn't agree with my being there.

A chickaree, a territorial squirrel of the Rockies, starts bombarding me with small pinecones from a tree branch above my head. Eventually, we come to a temporary truce.

I sit quietly while gazing out to the nearby river.

Every day it grows wider and deeper, more frenzied and forceful, as spring rains add to the melting snows that perpetuate this precious resource. The surrounding willows are waking to the new season. Their young branches of yellow and gold emerge from limbs toughened, grayed, and browned with age. The sprouting of their tiny buds soon will spawn intricate, lace-patterned leaves.

Willows provide filtered shade for other flora, resting places for birds and insects, and shelter for various fauna. They endow beavers, the engineers of the animal world, with materials to build or mend their dams and lodges. Beavers cut down willow branches with their strong, ever-growing incisors, and then drag them to their construction sites. Willows are, as well, a valuable food resource for them and also for other animals, especially moose that co-inhabit this riparian river course.

Finally, the rain abates.

I move to my vehicle, then travel to my own den for nourishment and rest.

* * *

Mother Otter gets few breaks during this time of year. While her offspring sleep, she must seek nourishment for herself and, in turn, for her young. Because she enters her oestrous cycle during the eight weeks following parturition, some of those excursions likely included reproductive activities that will lead to a new generation of pups next spring. That period now completed, her current feeding excursions are able to include some playtime. But her young await her return, slowly awakening from their sleep and anticipating their early morning meal.

Although the young otters' teeth have now emerged, they continue to be nourished by their mother's milk, gradually supplemented by increasing amounts of solid foods. As Mother Otter approaches the den carrying morsels of fish for her pups, the awakening youngsters rush to greet her with "otter kisses." Their hunger is very apparent. Often, she brings fresh-water shellfish and crustaceans for them to exercise their teeth and jaws—becoming toys to play with.

Outside their natal den, the sun begins to rise, dawning a new day. Mother Otter enters the outside world first, using her senses to detect any dangers. None perceived, she calls for her little ones to follow her. They both immediately respond, sensing that new discoveries await them today.

The otters' curiosity heightens as they begin to explore small creatures crawling in the exposed areas of Mother Nature's flesh. Their facial whiskers, "vibrissae," are very sensitive to the movement of external objects. The little otters try them close up as they brush their faces against various natural objects. This must give them an unusual sensation as they twitch their small noses. Vibrissae are one of the key elements they will eventually use in detecting prey underwater.

Their excellent manual dexterity is becoming increasingly evident, too, as their maturation process continues. This will enable them to hold onto their captured prey. For now, however, the young otters try using their manipulative powers to play with small pine cones and pebbles. They swipe their paws through a patch of snow that still coats their terrain.

Oh, there are so many new sensations.

Enough exploration for now; it's time for sibling play, rolling and tumbling and chasing each other, while not straying too far from their mother.

What prompts a bluebird to fly down and almost land on the little otters? Maybe to pluck a few downy hairs from their soft fur for its nest?

Whatever the reason, it ends the play bout and the three return to their den.

CHAPTER THREE

The Aquatic World

Swirling waters, what do you bring?
Otter and Beaver for us to cling?
Not all are pleased to see them, though;
but let's hope they will never go . . .

The Aquatic World

Mother Nature's plan for her faunal species is layered complexity, interwoven into an intricate pattern of life. Each species has its own purpose for being. The predatory aspect of the natural world may sometimes seem cruel to the human observer.

Is it any more cruel than that of Homo sapiens?

Mother Nature set her plans into motion millions of years ago. Adaptations of her life forms, along with the appearance of new ones, appear throughout the geologic timescale of Earth, each species finding its own niche. This evolutionary process is affected by many elements, including the perpetual changes within the physical world. Trying to unravel this complex web of life is the greatest intellectual challenge for humans.

We will neither totally understand its complexity nor will we ever totally control it. This is part of the plan. Let us hope humankind eventually learns to respect rather than destroy the components of this complex, biodiverse natural world. The intricacies of its fragile balance hold the key to our own survival.

One species in this biodiverse faunal scheme that plays an important role in the otters' world is the beaver. The activities of the beaver often equate to good otter habitat.

Riparian vegetation adjacent to waterways is an important cover component for the terrestrial habitat of our semi-aquatic otters. Beavers enrich this flora system through behaviors that can both increase or decrease local water tables. Beavers create good terrestrial and aquatic environments for a variety of other species as well, and are an important component in the health of this complex ecosystem.

Beavers are the engineers of the animal world. They are the designers, the contractors, and the builders of their own natural environment. They do it all. Their construction tool set is predominated by their strong front incisors. These teeth enable them to fell trees, chew off branches, and strip their bark. Beavers use the products of their labor for building dams and lodges, and for feeding. In fact, for their own survival, beavers must regularly use their continually growing incisors; otherwise, they would eventually curl back into their skulls.

* * *

Some Native American tribes ascribe attributes of persistence, hard work, and steadfastness to the beaver. They consider beavers as the "builders of dreams." I feel a particular connection to the beaver, perhaps because we share the same attributes. Or maybe it's because they permit me to share a special experience with them:

> *Encountering a beaver lodge while trekking in my study site, I sit next to it for some rest. Emerging from the little home are soft intermittent whine-like sounds. I sit quietly while listening to the expressions from one of Mother Nature's children.*

My imagination carries me into their home environment. Sitting motionless within the cozy encasement, I desire to enter the conversation with the little furry ones when an adult emits a loud gruff sound. This ends their vocalizing.

I softly apologize for my intrusion, then quietly walk away.

Birth of these young beavers occurred at about the same time of year as our otters. My imagination doesn't allow me to count the number of young conversing within the lodge, but there may be as many as four siblings. They are born fully furred. Unlike the otters, however, their eyes are partially opened. They will begin diving and swimming around in the water at four to five weeks of age, but will not be able to stay submerged until they are eight weeks old. At that time, they'll keep their fur water-repellent by grooming and shaking.

Sounds like a beaver dance! Ah, maybe that is the reason for their vocalizing and activity inside the lodge.

Beavers spend their first year of life within their close-knit family group, which includes their mother, father, and the offspring of the previous year. They reach sexual maturity and disperse from their home environment when about two years of age. The young ones I encounter still have several months to play, learn, and refine their different skills. This isn't unlike young otters, except otters usually disperse from their single parent at about one year of age, even though they won't reach sexual maturity until they are two. They have much to learn during their first year of life!

At some time, our otter family and the beaver family may actually spend the night at the same lodge. It's known to take place, but it sounds too crowded for comfort.

At least for now, our otter family remains in its natal area and where there aren't any beavers.

* * *

Today, a new adventure awaits the young otters as they emerge from their natal den. Mother Otter leads her two offspring into the forest. They've become used to the ambiance leading from the den into this pristine environment, spending many hours playing and exploring in this comfortable world.

What they do not know is that a new lesson awaits them today.

Following their mother, the calm waters of the pond lie ahead. They've seen this blue mass from a distance, but they never before approached it. This time is different.

When they reach the water's edge, Mother Otter doesn't stop. She continues into this new milieu. They start to follow her; but when their paws hit the water, they draw back. Although they are a semi-aquatic species, the young do not take readily to the water. It does not appear to be an instinctive behavior.

Mother Otter continues farther from them, splashing and playing in the aquatic world. They watch in amazement. She moves closer to the shore and splashes water towards them. They move back. She again swims to the middle of the pond playing, then disappears and reappears. They watch her. She returns to shore vocalizing her reassuring sounds as she swims back and forth from the shore to the middle of the pond. They do not follow her.

After her predetermined period of time, she goes ashore and picks up the little female by the scruff of the neck and drops her into the water. The male has always been a part of the threesome, so he follows them.

Cries ring out from the mouths of babes.

They quickly dog paddle back to shore, then move onto land. Their mother joins them for a bout of play. Once again, they seem to feel comfortable.

After just a few days, the young otters become used to the aquatic world. They now just follow Mother Otter right into the water. Their small bodies bob around like giant corks on the water's surface, allowing their mother to dive beneath them. During these sunny mornings, their dark shapes float atop the blue water, surrounded by reflections of trees swaying in the gentle breezes and mirrored snow-capped mountain peaks in this halcyon scene from the Rockies.

Sometimes, they chase unreal shapes in this looking-glass world. Other times, they chase real shapes that look like something their mother brings to them for nourishment. These figures jump above the water's surface, then quickly disappear below. The young otters try to follow them underneath the blue.

The male, however, begins seeing something different: an image to which he can't relate from his experiences. When he swims from his mother and sister to chase this image, a warning call sends him rushing back to them.

* * *

I enjoy observing the early morning swimming lessons of the young otters, but I must balance documenting their activities with my survey trekking. Now late June, the coolness of the mornings loses its battle to the heat of the day and pesky mosquitos play with my being.

"Whoops . . ." Another beaver hole nicely camouflaged by the thick willows and tall grass springs up along my path.

It's a good thing the young otters are out of hearing range or my vocal expression may have frightened them. My shoe sticks in the gooey stuff. With difficulty, I pull my wet, muddy leg out of the muck

The moment of minor distress now passed, my inquisitive nature sets in as I examine the hole carved out by the beavers. Beavers use these holes to come up onto land for feeding and for cutting off branches or felling trees. Once this portion of their job is completed, they slide Nature's woody creations into the water and move them to their construction sites or to their winter food stashes. The little otters will soon learn that they, too, can move back and forth from land to water through these holes. Unlike the beaver, however, they don't eat the flora or build dams and lodges.

"Okay, so the hole has a purpose," I laugh while trying to scrape some of the mud from my leg. After all, this is just a minor inconvenience en route to my goal of contributing information towards the otters' survival.

Squish-squish . . . a-trekking, I will go.

* * *

The young otters still have several weeks before they become efficient swimmers. "Practice makes perfect" will become their motto.

During their early stages of swimming, the little ones' movements appear clumsy when compared to the fluid motions of their mother. Her grace and beauty moving through the water is like a bird gliding through the skies. The otter's speed, placed at 6 to 7 miles per hour, is combined with agility in catching its prey. A fish can have a hard time outmaneuvering this body of flexible motion. Our little otters will need much practice to reach the swimming abilities of their mother. It is a part of their survival.

The streamlined flow of our otter is well adapted to the aquatic side of its life. The otter's head is small and flattened, widening into a thick neck and cylinder-shaped body, ending in a relatively long, pointed tail. Its small eyes and ears are set high on its head on a plane with its bulbous nose. This configuration allows the otter to swim low in the water and still use all its senses.

The otter's legs are short and stocky with inter-digital webbing between all five toes. Otters actually use a variety of swimming techniques involving their forelimbs, hind limbs, and tail. Each stroke cycle consists of a power phase and a recovery phase. During the undulating, or porpoising movement, the otter flexes its body and tail while paddling with its hind limbs.

When returning to their den after swimming lessons and play behavior, our babes are both hungry and tired. After Mother Otter feeds them her milk, supplemented by fish she caught for them, they enter their sleep world.

Since animals do dream, some of our young otters' dreams may be about the new world they are experiencing. Maybe they dream about the real and unreal shapes they chase in their looking-glass world. Maybe they dream about the playmates that wouldn't play with them, but instead dive or swim away.

The male otter, however, enters into a different dream world. He sees a small white shape, fluttering and dancing in joyful glee. This image looks like the one he started chasing in the aquatic world. It gives him a feeling of warmth and happiness. He wants to join it; it's having fun.

Then, it disappears!

He awakens and looks for it, but it's nowhere to be found. Oh, well, maybe he'll find it during his next aquatic adventure. He secretly hopes that will happen as he returns to sleep.

* * *

I, too, am tired and dirty, so I return to my den for a shower, some food, and rest. Prevailing clouds dominate the skies. Thunderstorms are customary at this time of year in the Rockies.

I walk into my cabin.

Shadows dance across the cabin walls as the sun darts in and out of the clouds. The late afternoon sun eventually loses its battle and darkness takes over.

I shower, work on my data, then eat a meal. When my work is completed, I climb up the steep ladder-like staircase to the loft of my cabin and to bed. I desire sleep to come easily tonight, since otter trekking begins before sunrise. I fold my head into the soft pillow, then fall asleep. Unfortunately, it doesn't last.

I awaken with an abrupt start, trembling uncontrollably.

Outside my den, strong winds blow through the darkened world, resonating like a pack of wolves howling their expressions of dominance, reinforced by a thunderous applause. Hail begins to hit the roof, resembling busy woodpeckers keeping the beat. With their night in the spotlight, streaks of lightning create dancing shadows.

Wrapping within my cradle of down, I try to calm myself.

My trembling subsides as I remember what awakened me. I dreamt a small white shape fluttered and danced in a state of joyful glee. It was frightful, yet beautiful.

What is it?

I must objectify this image. It seemed so real! Maybe I'll figure it out in the morning. As sleep returns, so then does my mysterious dream.

Soon, I will come to understand that this is the first of many encounters with the spirit of the young male otter . . . powered through the great Otter Spirit.

CHAPTER FOUR

Worldly—Unworldly

Our spirits dance in childlike play,
while Earthly forms work hard all day.
Much to teach each other's being;
but for now, this is appealing . . .

Worldly—Unworldly

The otter's spirit and my spirit regularly enter the dream world of each other's Earthly forms. When not educating, our spirits sometimes meet just to frolic. Lifted into another dimension, these wisps of white turn into beautiful clouds sculptured in our worldly profiles. Here, we engage in bouts of play while chasing each other across the blue, ducking behind other billows of white in games of hide-and-seek.

When it's time to end our amusement, the great Otter Spirit absorbs us into One.

This immersion lasts only a short time, but it's an important step in connecting us closer together. When we again separate, we return to Earth to continue our respective teachings.

I feel a more formidable bond with my special one.

* * *

Last night's rains were a necessary cleansing of Mother Nature's embodiment. Precipitation brings purification of her being and of her soul. I also need such a cleansing.

When ambient temperatures are warm and the rains gentle, I dance within their purifying embrace. When temperatures are cold and the rains forceful, I observe respectfully from a natural shelter. Whatever the weather, my internal cleansing comes from being within the natural world. This is where I sense that which is important to my worldly form.

Yes, nature purifies my soul.

How does the otter cleanse his soul?

Humans may never know that answer. But how he cleanses his body is essential to his very survival. Although the rain's freshness washes him down, more important cleansing comes from immersion in the waters of his aquatic world, followed by careful grooming of his fur. As the seasons progress, the otter will learn that this aquatic environment is not only an important part of his life during warm temperatures, but during the cold as well.

But the otter doesn't have to worry about becoming cold. His species evolved with densely packed fur. His layered coat consists of stout outer guard hairs which form the waterproofing quality of his fur and finer inner hairs which form the body-warming quality. The density of fur in the otter's mid-back section is about 373,000 hairs per square inch!

While he lacks a layer of fat, unlike many mammals who live predominantly in the water, he actually uses air for thermo-insulation. During a dive, air gets trapped within his densely packed underfur which, in turn, insulates his body from the cold waters. For this air-trapping capability to be effective, his fur must be kept clean and in good condition. This is why both the young male and his mother regularly groom his fur and why he rolls and rubs on the terrain of his habitat. He senses these behaviors are important to dry and maintain his fur.

What is grooming? I groom my hair by using a hairbrush. The otter grooms his hair by pulling it through his front incisor teeth or by drawing it through the claws of his digits, using them like a comb. These behaviors, in combination with rolling and rubbing his body on the terrain, help keep his fur healthy and dry and, in turn, keep his body warm.

Some river otters living in other parts of North America, as well as close relatives that reside in other parts of the world, are also fresh water otters, but they colonize the saline environment. Although they spend time swimming and fishing in salt water, researchers discovered that an important part of their coastal habitat is the containment of some fresh water accumulation. Fresh water is necessary for them to wash salt from their fur to retain its thermo-insulating properties so essential to their survival. (At least, our otters don't have to worry about removing salt from their fur.)

Researchers in Scotland have observed that some of our otters' relatives dry themselves in another manner. (No, they don't rub themselves down with their kilts!) After a nice fresh-water cleansing, they move into their dens and come back out again through a narrow entryway. As they pull themselves through this tight passage, the water is squeezed from their fur. They then continue their activities in a nice clean, dry state. (Maybe our otter should try that sometime, as long as he doesn't get stuck!)

* * *

The comfort and security offered by their natal den is about to change for our otter family. Mother Otter decides it is time for her young to meet the Colorado River, where they will spend much of their lives. Now about twelve weeks of age, the otters are better able to travel with their mother than when they were younger.

Under the cover of predawn, the little otters follow their mother into the yet unknown. Mother Otter is well acquainted with the waterways and potential den sites in her chosen living space. Therefore, she moves the family to an area not too far away from the natal residence, thus enabling her young a shorter distance to travel.

Following the course of their inlet, the family reaches the Colorado River. The little otters look out onto the swift-moving flow. They may secretly desire to return to the tranquil serenity of their pond as they draw back from the water's embrace.

The threesome do not directly enter these waters, but instead move along the shore. Up ahead, they see another pond of calm waters and the youngsters feel more comfortable again.

This pond was created from waters backed up behind a well-constructed dam and removed from the river's flow, thanks to their benefactor, the beaver. The main purpose of the dam—constructed of branches, mud, and rocks— is to hold the water level constant in the pond where the beavers' lodges are built.

That beavers are still active in the area of the otters' new environment is evident from signs of recent activities. However, they moved to a newer lodge. So, again thanks to the beavers, the otters may now use one of their benefactors' homes. Because our otter family will not do its own construction, they can't be too choosy.

The otters enter their new home. The foundation and walls of the lodge are constructed of branches and mud

with a dome of loosely placed boughs so air can come through for ventilation. The floor is above the water level, but the lodge has two underwater entrances into the aquatic world of the beaver pond. The pond contains a good prey base for the otters' diet, with thanks again to the beavers' engineering. In this environment, our young ones can begin to learn a most important yet difficult task: how to fish.

To paraphrase an old saying: "If you give an otter a fish, he'll eat for a day; if you teach him to fish, he'll eat for a lifetime."

The young otters are now almost fully weaned and ready to learn their new skill. Although they began eating solid foods shortly after their emergence from the natal den into the outside world, they continued to be supplemented by their mother's milk. This will taper off during the next two weeks, but comfort suckling may extend until they are five months of age.

The little otters explore their new surroundings. This home is larger than the one they just left and contains some interesting nooks and crannies to whet their curiosity. Their sense of smell detects unfamiliar odors. Mysteries abound in their new world.

Eventually, their activities cease and the furry young ones fall asleep. Outside their world, a dark shade is pulled across the sky. Soon Mother Otter will leave to search for food for both herself and her babes. But until then, she joins her offspring in a nice nap.

A strange thumping sound awakens the family!

They cautiously peek out from the above-water dome. An unusual looking creature stands beside the lodge. Unconcerned, Mother Otter draws back into their home, giving comfort to the little ones that their intruder isn't a threat, but not curtailing their curiosity at what they see.

The young otters will meet a more diverse group of animals in their new locale than those few they met in their natal area. Predatory animals will be a definite concern. Not so for the others, though, as long as the otters don't get too close to the foot of a moose or the spines of a porcupine.

Oh, they have so much to learn.

Various types of vegetation surround the otters' new home, including the willow. The strange animal seen by the otters happens to be a moose, an herbivore that feeds particularly on willows. This animal feeds not only on willows but also on aquatic vegetation; both types of plant life are influenced by beavers' behaviors. Although, in the Rockies, moose typically won't spend their entire year in these wetland environments, the nutrients they receive when they do inhabit them are an important part of their diet.

Early the next morning, a heavy fog covers the beaver pond. It is so thick that no one from the faunal world can see much past the nose on its face. Mystery and intrigue surround the otters' world. The young ones have been swimming and diving in the pond with their mother since before first light, so they don't appear too concerned with the gray curtain drawn across their world. At least, that is the case until something large and unexpected suddenly appears before them.

Unable to immediately determine what is emerging through the thick fog, Mother Otter vocalizes a warning, sending the three diving to the underwater entrance to their home. The astute mother knows it is probably a moose, from its size and general shape . . . but it could also be a bear, which would be a potential threat.

Mother Otter remains vigilant as the intruder stalks their lodge, nibbling at its willow walls.

Soon, the intruder walks away, convincing her that it is a moose and allowing her to relax her guard. The little ones curl within their mother's warm embrace and seek comfort suckling.

* * *

I empathize with the otters.

This morning, I slowly inch my way through the fog towards the river. Turning back to my cabin is a definite option; but since nature's mysteries abound in this atmospheric condition, I continue. Plus, the gray curtain will eventually lift.

I step out of my vehicle and carefully walk towards the river. The morning mist that fills the air covers the ground with a silvery hue. I feel the moisture against my face. How lucky to experience fog on the river.

As the fog thickens, the reality of my situation sets in. I feel encased within an ugly, gray, barren world. My senses are muted and I feel disconnected from Mother Nature.

Has she deserted me or enveloped me?

I try to imagine a world without Mother Nature and tears begin flowing from deep within me.

No, such a world must never occur!

Trembling, I grasp onto my medicine bag for assurance.

Nature's world again becomes evident as I begin emerging from the fog. I start slowly walking. I slip into another beaver-created hole. No complaint this time as I pull out my wet, muddy foot. After all, one of her children worked hard to create this hole.

Dirty and wet, I sit down next to a large tree that offers a little protection from what might emerge from the fog, pull a thermos from my backpack, and pour a cup of coffee. I remember why I chose this particular location. The small white shape, which entered last night's dream world, guided me here.

Yes, I am near the beaver pond of our otter family.

Out of the gray willowy world, a large figure emerges. Slowly, I slink to the opposite side of the tree. I desire to run, but that is not really an option. As long as I keep that undefined space called "flight distance" between the animal and me, I might be safe. If I inadvertently enter into that space, the animal might feel threatened and attack.

I wait.

The fog begins to lift to the point that I now see the shape of the animal. Yes, it is a moose!

This is probably the same animal that earlier frightened our otter family. While not a real threat to them, its sudden appearance, emerging from the gray mist, is awesome and intimidating. The major concern with the moose is its size and unpredictability, particularly if encountering a female and her young, or a male during rutting season.

Realistically, no wild animal should be approached too closely by a human. Although sometimes unavoidable, it shouldn't be a deliberate act. They should be quietly observed from a safe distance.

I hold a deep reverence for the natural world. I tread lightly across the land. Because I'm intruding into the homes of the various wildlife species, I feel guilty disturbing any of them while trekking within my study area.

I apologize and slowly walk away.

* * *

After their frightening experience, the otter family takes a nap.

The white image again appears in the male's dream world, and the spirit shows him the animal that frightened his family. It looks like the same animal he saw from the lodge on the previous day. It is a moose. The animal is feeding on willows next to the beaver pond.

Then, a smaller version comes into view jumping, running, and playing. This young animal was also born in the spring. It weighs 50 to 60 times the otter's weight at birth, and is already following and browsing with its mother only two to three weeks after naissance. However, it will not be completely weaned until about five months of age.

When the male otter awakens, he wants to play with the young moose calf, despite that it's bigger than he. Peeking out from the above-water dome, he sees neither the mother or her calf. Confused between the visions of his dream and that of reality, he comes back down into the lodge.

The real moose probably moved away under the cover of gray, since I don't see it either.

Mother Nature continues to lift her misty curtain, revealing simple sculptured forms. Her world appears as beautiful simplicity, yet is realistically complex.

I rise and walk slowly to the nearby beaver pond. Concealing myself in some vegetation, I await my research animal.

After all, the otter's spirit drew me to this location.

CHAPTER FIVE

Lessons

To chase fish is so much fun,
but to catch them must be done;
pups don't use a pole and line,
but their method works just fine . . .

Lessons

Sitting quietly while awaiting the otter family's appearance at the beaver pond, I witness the emergence of life awakening from the fog. Cascading waters course over the beavers' dam in contrast to the quiet shimmering waters of the pond it shelters.

Enter the kingfisher with its rattling, sewing-machine-like call and hurried flight, stopping to perch on top of a willow. Here, it awaits an unsuspecting fish to jump above the surface in pursuit of its own meal.

There's one now!

A quick plunge headfirst into the pond and the kingfisher emerges successful.

Wow! One catch and two species: fish and ingested insect.

Swallows circle the pond making swift dips into the water to catch water-borne bugs, while also adding flying insects to their diet. There goes mama duck followed closely by her five little ones. The ducklings paddle quickly to keep up with their protective mom, each looking like a child's small wind-up toy.

A hawk suddenly appears, soaring quietly above the duck clan. It is one of several predators that can strike swiftly from the skies above or from the shadows below. Its appearance can strike fear into a hapless family of swimmers. But this time, the hawk passes over in search of a meal elsewhere, a threat no more to our little ducklings.

Melodic songs emerge from tiny shapes perched atop the bushes. Grateful that the hawk chose not to shop here for a meal, the little birds sing their joy.

To our ears, these bird songs are beautiful music; to the birds, they have specific meanings. Each species has its own sound repertoire, with specific songs sung only during certain seasons and with particular behaviors. There can even be geographic variations within the same species. How complex Nature is, all with a purpose.

While listening to the music from these small wonders, beautifully colored butterflies gracefully flutter through this scene. Nature definitely sprayed her love potion over the pond this morning.

Hearing the birds and watching silent butterflies remind me of a legend from the Papago Tribe of southwestern Arizona, telling of the time when songbirds actually lost their music to the butterflies at the hand of the Creator.

I watch as the array of colors from Mother Nature's garment flow across her body. Are these the same hues gathered by the Creator in the legend when He made butterflies? These small quiet jewels gladden my heart. I stand to join the butterflies and, too, flutter gracefully to the flowers, soaking in sweet nectar among my newfound friends.

Then, the butterflies start singing to me while the songbirds observe in silence.

No, this is not right!

The songs have been taken from the avians. The Creator gave butterflies the colors of the rainbow but He also gave them the songs from the birds. Fortunately, according to the legend, He realized His mistake and returned them their music. Now, all can listen to the avian melodies and watch the butterflies' flickering colors.

Where are the otters?

Savoring Nature's wonders, I almost forget the reason I am here. It's already late morning. At this time of year, I know that otters are typically more active during the crepuscular hours; yet there have been sightings at all hours of the day, particularly in more secluded areas.

Already getting hot and buggy, I decide to leave and return near dusk when the otters will likely be active.

After all, the little spirit guided me to this spot.

* * *

While I patiently awaited the otter family, they were resting and exploring their new home environment. They had already been active chasing fish during the early hours of dawn. Mother Otter did more than chase, however; she caught them all breakfast. Under cover of low light, the young otters followed their mother onto the shore where she laid before them a large fish for their nourishment.

During the early months of her offspring's life, Mother Otter brings them their food. She catches larger fish for them than those she catches just for herself. Although the young otters are chasing fish at this stage in their lives, they still need several more weeks of training before becoming successful at catching one. They have a lot to learn.

Otters are opportunistic feeders. While they consume prey that supplies adequate caloric content, they target those that require a minimal expenditure of energy to catch. Because their overall diet consists primarily of fish, where there is a choice, they will select slower-moving species.

Slower-moving fish generally include forage or non-game species. The otters' removal of these competing and less desirable fish (less desirable for human consumption, I can attest) actually benefits the game-fish population. Our otter family is doing a good deed for humans.

In addition to catchability, many factors enter into the otter diet equation. Our young otters must learn about different aquatic habitats, prey availability, seasonal variations, water characteristics, effects of time-of-day, and inter- or intra-species competition. Yes, these are the reasons our young otters need such an extended period of time for their training process. They have to learn how to catch their prey, where and when to catch it, and how to avoid competition.

* * *

Dusk arrives.

I return to await the emergence of my animals. The sun disappears behind the majestic rocky peaks, lending colored brilliance to the sky and spectacular reflections to the calm waters of the beaver pond.

There is a disturbance in the waters!

Is it possibly an otter?

No, the swimming behavior tells me it is a beaver. I spot another with a branch in its mouth, and then another. They have arrived to work on their dam. As the evening progresses, their behavior suggests that something unusual is happening.

I wonder how they can bring the branches to the dam so quickly.

I slowly rise and walk upstream along the pond until reaching what appears to be an older and partially destroyed dam. I sit down and watch.

The beavers are taking the building materials from the old dam and moving them downstream to the construction site of the newer one.

Yes, the beavers are recycling! Nature's engineers are also one of Nature's eco-sensitive species! What a fascinating observation.

Ever persistent my beaver be,
clever recycler of Nature's tree.

Building dams is instinctive for beavers. Because beavers are vulnerable on land, the deep water behind their dams affords them a safety net while foraging for

food and building materials. Another benefit is the assurance of access to their stores of food submerged under winter ice near their lodges.

Recycling, however, must be learned. Interestingly, this behavior goes further to substantiate a scientific fact in the animal world. The beaver's recycling behavior and the otter's feeding behavior allow each species to reach its respective goal with the least amount of energy expenditure.

So many layers and complexities exist for each species of the natural world.

Under the cover of darkness, I return to my den. Disappointed at not seeing the otters, I nevertheless feel fortunate to have witnessed one of Nature's secrets: recycling beavers.

Climbing into my cradle of down, sleep comes immediately to my exhausted being.

* * *

The otter's spirit enters my dream world.

I was looking in the wrong location. While sitting near the large dam, the otter family was moving through one of the waterways extending out from the pond.

Over the years, beavers developed a complex and extensive system of channels that left many islands among the waterways. Although there are pockets of shallow waters, the channels contain good water depth.

Our spirits grasp onto the wings of a bird and look down upon the entire channel system. It looks like a giant octopus with its arms extended above the watery world. This makes it impossible for non-aquatic species, like real me, to penetrate the habitat. Even a canoe cannot navigate this system.

Just think of the protective value it affords the lives of otters, beavers, and other species of fauna who reside here. Various species of fish live and reproduce among the channels. Mother Otter knows about the habits of such aquatic life. It's all part of Nature's balance.

This habitat ensures a perfect world for Mother Otter to teach her pups some of the lessons they need to survive in the challenging world they must eventually enter.

* * *

During the weeks through July and August, I observe some lessons taught the young otters involving how, when, and where to forage. Those behaviors, hidden from my view, are shared by the otter's spirit during my dreams.

The pups' undertook their first excursions when they were about fifteen weeks of age, but they didn't catch their first fish until almost twenty weeks of age. This later date corresponds to the time they reached their adult swimming proficiency.

One of the methods Mother Otter uses to teach her pups to catch fish is to provide them with a live fish. Sometimes, she drops it in one of the small shallow ponds of the beaver's system; other times, she'll carry it onto land near the water's edge. Because the fish is still alive, it makes for an energetic chase for the furry masses of motion.

Their antics appear uncoordinated and comical, similar to those of a young dog at play. The otters consider this play like chasing a live ball. In the early stages of this lesson, the fish win the game; over time, however, the fish will be no match, and the otters will win their just reward.

An advanced instructional technique encourages the young ones to follow Mother Otter on her fishing expeditions. She swims swiftly through the large pond followed by the pups, trying to imitate her speed and agility.

Sometimes, my spirit joins the family, creating confusion for the male.

Am I a fish or the white image that enters his dream world?

When he chases me instead of his mother, he loses an important part of his training. I eventually retreat. But to real me, allowing my spirit to become an integral part of his learning experiences deepens the bond of our two beings.

The young otters must learn not only how to catch fish, but where to catch them. As they follow their mother on her foraging excursions, sometimes they explore weedy areas where certain fish and other aquatic species hide from predators. At other times, they explore bottom waters where they find slow-swimming species such as suckers or the more sedentary sculpin. At the other end of the spectrum, the otters discover species from the salmon family, which include both salmon and trout. The young ones soon realize these species' superior swimming and hiding ability give them less return on their expenditure of foraging energy. (Remember the scientific fact substantiated by the otters' fishing behavior?)

Where they occur, another favorite food of otters are crayfish. This crustacean becomes a major part of our otters' diet during summer months, although they can possibly consume them at other times of the year as well. Crayfish, usually not found at a depth below three feet, become particularly accessible to otters. This doesn't mean they can always be easily found. Often, crayfish hide in vegetation or under small stones that the otters must overturn with their muzzles to reach the tasty morsel.

The otters' mother will continue to provide them with food until they become proficient hunters. That expertise may take nine or more months.

Otters catch fish in their mouths. Their teeth are designed for grasping, grinding, and crushing; their carnassials are used to shear the flesh. They hold the fish with their hands, eating it beginning with the head to ensure completion of the kill. Normally, they devour the entire fish. They crush crayfish in their mouths, then consume the tail or thorax. The size of fish that otters catch varies, but it is usually up to twelve inches. They eat smaller fish in the water, but take larger ones onto shore to enjoy their feast.

Otters can stay under water for two to three minutes before surfacing for life-sustaining air. Their foraging dives normally last less than one minute, usually about twenty seconds. The duration of their dives depends on prey availability and habitat. Seeking out prey that are hiding in weedy areas can take longer than stalking moving prey in open waters. Of course, it all depends on the type of prey they seek.

Although they can dive to a depth of about 60 feet, most river otters dive less than 10 feet while hunting. The otters' ears and nostrils close off during submersion. They use their vision to find prey on bright days and in clear waters. They use their vibrissae to assist them with hunting in murky waters or on dark nights or in winter when available light is attenuated by the thick ice. These sensitive whiskers can detect the very slightest nearby movements. Our young otters made that discovery after emergence from their natal den. However, even under the most ideal conditions, not all their dives are successful.

Our young otters learned many important lessons while residing in this beaver-created system. But now, Mother Otter decides it is time to move the family onward. There is still more to learn and more to explore as their journey of life continues.

I, of course, will join them as the male's spirit and mine become more deeply intertwined.

CHAPTER SIX

Explorations

Sounds abound during this time of year,
*   springing forth from male species of deer.*
Nature prepares her palette of color
*   to change from one season to another . . .*

Explorations

*I*nto their sixth month of life, the young otters enter my favorite time of year—autumn. Their excursions during this special season will bring them into contact with new animals and new sounds. While Mother Nature busily paints the leaves of her garment, the mature members of her deer family begin their rituals of rut.

The dominant males of the three deer species residing in the Colorado Rocky Mountains—elk (Wapiti to many Native Americans), mule deer, and moose—strut their brawny bodies across the lands, holding their heads high to display their impressive racks of antlers. Sometimes these animals adorn their antlers with vegetation, elevating themselves to proud knights of Mother Nature's castle. Yearling males step aside for the older, handsome members of their kind. The young stags' turn will come, but it will be another two to four years before they enter the onset of their prime.

The deer's musical composition begins with the low moaning calls of moose cows and the bellowing crescendo of aroused bulls. Then follows the haunting melody of bugling male elk. The drama that unfolds as the voices of autumn leap from the pages of this ancient musical score never gets old, never gets stale, never tires.

* * *

The otter family moves through portions of the female's chosen land mass. These excursions add to the knowledge of the young otters. They learn new places to search for different varieties of prey. They become familiar with various types of den sites. They experience new visual, aural, and tactile events. They meet new species of wildlife.

One of their lodgings is near the home of another semi-aquatic species related to their beaver benefactor, except it is smaller and with a different looking tail. The two young otters start to chase the animal through the water, but it disappears under the blue. They dive but, not finding it, emerge after a few moments.

This new animal is a muskrat, which is about one-third the size of the beaver. Like the beaver, it can stay underwater for fifteen minutes, in contrast to the otter's maximum duration of two to three minutes.

Maybe someday, they will find another species with whom to frolic. In the meantime, they play with each other. They dive into the water, coming together and tumbling. First, one pops above the water, then the other. They resemble a Ferris wheel rapidly rotating through the water.

What fun! They discovered a new game.

But after a few minutes, Mother Otter decides it's time for a rest before moving onward.

One evening in early October, the otter family slowly swims along a river segment coursing its way through a large grassy meadow. Astonishment would best describe the young otters' expressions while they move through a performance of rutting elk. Haunting bugles and clashing antlers resound throughout their world. Add to this, the numbers and size of these majestic beings. It's just too much to comprehend for our babes.

The large handsome males look like giants to the small otters. After all, these elk weigh up to 40 times their body weight. Even as adults, otters average only twenty pounds. They stare at the large protuberances extending from the deer's heads. The spread of these antlers is longer than the total length of our otters.

Yes, the elk are an impressive sight.

Although curious, the young otters' fear is apparent.

The family continues onward to their "shelter du jour" in a heightened state of alert. The particular den they select for the evening resides within the bank of the river, once again thanks to the beaver. It is nicely carved out of the soil of the embankment, with one entrance above the water level and another one below. No beaver family is staying there at the time, so the otters take advantage of the "vacancy" sign.

Tired, the three otters curl together into an otter ball and quickly fall asleep.

* * *

Once again comes my little white spirit entering the male otter's dream world.

He finds himself back again with those large noisy animals. They pay no attention to him, as if he's not at all there.

The elk, an herbivore, feeds on grasses and other plant products. They are a gregarious species with differing herd compositions, depending on the season of the year. The only time an elk would present a threat to an otter is if the otter gets in the path of a focused animal.

My spirit brings the attention of my special one to a particularly dominant bull. The little otter stirs in his sleep as he visualizes this robust giant prancing about the meadow's stage, proudly bearing his majestic antlers, intimidating all those around him.

During a four- to six-week period, usually from mid-September through October, big bodied, dominant, mature, well-racked bulls exert all their stored-up sexual energies during this most lustful season. These bulls gather, guard, and hold a harem of up to thirty cows and their calves—all while warding off other male challengers through a combination of displays and occasional head-on clashes. They must mate with the mature females if they are going to be successful progenitors. In such an agitated state, they bugle, thrash bushes with their antlers, roll in urine-soaked wallows, and test the females' receptivity through a behavior called "flehmen."

This is definitely elk season.

I never tire of watching this impressive performance on Nature's stage.

Now, through my spirit, I can share it with the otter male and teach him there is really nothing for him to fear. Perhaps the young otter secretly wishes he could become one of those big impressive bulls, but that is not a part of Nature's plan for him.

The spirit begins to fade from his presence and he returns to a deeper state of sleep.

* * *

The next day, the otter family is again on the move. Soon they find themselves in the arm of a waterway leading towards a beaver pond. Here, they encounter a different deer species, but one which they recognize. It appears much larger now and bigger than the giants they saw yesterday.

The moose is primarily a solitary animal, so their rutting behaviors differ from the harem-structured elk. Even though the otters meet only one moose, Mother Otter signals her offspring to be careful, particularly at this time of year.

Moose are the largest members of the deer family. A mature bull can weigh up to 90 times that of the otters. While imposing enough, add to this his palmate antlers that spread almost one and a half times the otters' length. His antlers alone can weigh over four times an otter's weight.

The family decides it might be best if they move to another area for now. The moose isn't a predator, but this rutting bull is a very focused animal and headed in their direction.

* * *

I too encounter moose during my trekking, but do everything in my power to keep a safe distance from them.

Early that same morning, I approach the large beaver pond diversion. The male otter's spirit again guides me to their possible location. Having no idea the otters have already been to and departed from the area, I walk out onto a small peninsula of land to look for them, or at least for their signs.

I happen to know this is a moose bedding area. Although a seasoned wildlife researcher, my otter quest sometimes clouds my common sense. This is one of those mornings.

While bending down to check some animal tracks, I hear low grunting vocalizations and vegetation rustling sounds. Slowly, I straighten my body, then turn around to see a large bull moose moving in my direction!

My heavy breathing and pounding heart throb audibly in my ears.

Will the moose also hear them?

Trembling, I stand as quietly as possible, while clutching my medicine bag.

I have only one obvious exit from this small land peninsula.

But before the moose approaches my exit, he turns and walks into nearby pines and willows.

I breathe a sigh of relief, then cautiously make a hasty retreat.

The female for whom the large bull is looking is probably the one I sometimes encounter in that area. The cow probably left behind some clues as to her sexual state. It is up to the bull to follow her trail and find her. When they do meet, their mating will take place during the course of approximately one week. Following their lustful period, the bull will leave and search for another receptive female.

When I again get back to the river, I look over towards the beaver pond. The bull moose has circled around and now stands on the small peninsula.

"It's all yours!" I tell Mr. Moose.

I resume my trek away from him as fast as possible.

* * *

While the male otter sleeps, my spirit shares this recent experience with him.

I advise him that he should be cautious when encountering bull moose during the rutting season. Although strictly an herbivore, a focused bull, during his heightened physiological state, may attack almost anything. Plus, moose are good swimmers. My shared experience reinforces the caution expressed by his mother.

The male otter moves closer to Mother Otter and curls up within her warm embrace.

* * *

The otter family continues its excursions in the river and its drainages. Sometimes, they move across land bridges to reach one waterway from another. It is during one of these land ventures that they encounter the third member of the deer family residing in our Rocky Mountain region, the mule deer.

Although not a real threat to otters, males can weigh up to 16 times an otter's adult weight. This species inhabits a variety of environments. Less solitary than the moose, mule deer are not as gregarious as the elk. They get their common name from their large ears.

The mule deer also ruts during autumn but, in some locations of our country, their rut extends into January. Mature males, known as bucks, attempt to assemble a small harem of three to four does, but are not known to actively keep them intact. Just courting one female at a time seems to be the norm.

Mother Otter decides to let her little ones watch this deer's behaviors before continuing onward, but to watch from a safe distance. Two mature bucks encounter one another. They engage in several ritualized displays, proceeding into a shoving match with their antlers intertwined. One of the bucks finally leaves the area.

The otters quietly observe the remaining deer's behavior. He rubs his facial scent gland against some vegetation. Then, he urinates into scrapes he dug underneath the marked flora. By performing these behaviors, the male attempts to attract a receptive female and ward off any other bucks that come into the area.

While the young otters watch these behaviors, they wonder how they will engage in mating when becoming sexually mature. For now, they can just watch these other species—at least until Mother Otter decides it is time to move onward.

The otters gaze up into the branches of deciduous trees. Colorful autumn leaves tremble with anticipation of what lies ahead as they slowly break loose from the arms which have held them since their emergence. The leaves appear to wave good-bye as they float across the landscape, encasing the otters in a familiar sensation. (Yes, this was the "enemy" that the male tried to bat away when he was just a baby. Ah, they aren't anything to be scared of.)

The otters survey the land, now containing shimmering frost-covered vegetation on this sunlit day. Even the brown grasses look like a diamond-studded carpet.

* * *

The shorter days and brisker temperatures send a warning to all forms of life that a change will soon come upon them. The period of rut has ended for the deer, so the focused males must now feed diligently to rebuild their strength for the upcoming season. Some of the smaller animals hastily gather their stashes of food in preparation for what lies ahead.

The continuously working beavers must do so at a quickened pace. A new world will confront their lives when ice covers their waterways. They must cache enough food for their winter survival needs. They re-examine their dams to ensure the larger, focused, rutting deer imparted no damage to them. They hasten to cement branches into the bottom of their ponds and close to their homes. They add fresh mud to their island lodges. This mud will freeze during the cold temperatures of winter, providing an impenetrable fortress from predators that can walk out to the lodges when ice covers their ponds.

This fast-approaching new season can be the most difficult period of the year for the various fauna. Most individuals will endure, but some will not. This is a part of the layered complexity of Nature's master plan.

The young otters, too, will engage in new challenges and learn new lessons. They don't migrate, hibernate, or stash food, so where they go and how they eat when snow and ice covers their den sites and food-rich waters will determine their survival.

Mother Nature's white crystal tears begin to fall upon the land. The little otters appear unconcerned about their survival as they bat with their paws at this new-found toy.

This is fun. But, they can't feel it. Where did it go? Oh well, it seems they've been provided with a never-ending toy supply. Maybe they can even learn to make snowballs!

Here comes winter . . .

CHAPTER SEVEN

The New Season

A carpet of white now enters our story;
Nature's new beauty brings forth all her glory.
The otters discover new games to play;
watch them slide down the hill of the day . . .

The New Season

The curtain slowly rises as the massive spotlight is lifted above the mountain backdrop. Freshly fallen snow shimmers like precious jewels spread across this sunlit stage. Glistening chandeliers, suspended from boughs on high, add their own decorative ambience to Nature's theater. Verdant pine needles peek out from their feathery white coats to watch as new scenes unfold during this winter presentation. Soaring across the cobalt blue sky, large ethereal sculptures release some of their soft white flakes onto the land.

Listen to the silence.

All seems at peace in this world.

Oh! There goes a raven flying quietly across the stage.

But where are the other faunal actors? Did I miss them?

Fortunately, they left behind clues to their particular portrayals.

Suddenly, the silence is broken!

I topple over into the snow. Concentrating so hard on clues left behind by the performers, I'm not watching my path of travel.

Ah yes, soft snow surrounding willows once again collapse underfoot and carry me deeper into the surrounding snowscape.

From my reclining position, I look up into the bare branches of a nearby tree where a raven sits observing my clown-like behavior.

"You could help me," I call out to the bird. "After all, you have magical powers."

The bird just sits quietly, observing my antics. Then, it flies away. I know the raven will return, but only he knows that proper timing.

"I will just have to use my own powers in this situation," I say to no one.

But am I now, and will I continue, to use only my powers in such situations? Or are Mother Nature and some of her children my guiding force? Was the raven's presence a foreshadowing of the final immersion of my spirit with the otter's? Or was he just a casual observer of my predicament?

I strongly sense the former.

One of Nature's children, in whom many Native American tribes hold a deep reverence, is Raven. They believe this bird carries the medicine of magic. Some tribes believe this species has the power to blend human and animal beings, as well as their spirits, into One.

Is that why I sometimes appear to be almost otter-like in my behavior? Is that why it is important for the otter's and my spirits to be absorbed into One? Has Raven actually used its magical powers on us?

Whatever the case, I sense Raven's return for a significant role in the convergence of our two spirits.

While I struggle to lift my snowshoes from the soft white down, the otter family moves effortlessly through their new terrain. The young otters discover that they have built-in snowshoes.

The inter-digital webbing between their toes, that assists with their aquatic maneuvers when swimming, confers a snowshoe-like quality on nature's blanket of white, particularly to their larger rear feet. Four small rough protuberances on the heel pads of their hind feet also provide them with greater traction on slippery surfaces, such as mud, snow, and ice. That the weight supported per unit of their foot surface is so much less than that of mine further enables them to stay atop loosely packed areas of snow through which I easily fall.

Finally getting back on my feet, I ponder how much easier it would be for me to be an otter.

Cold, I walk to a nearby tree and sit within Mother Nature's embrace. I reflect on the fun the otters have while sliding across this blanket of white, easily acclimating to the snow and ice of winter. Their multi-layered coats and adaptable behaviors keep them warm during this season of the year.

Closing my eyes, I draw deeper into my own heavy coat—and soul. I find myself sitting in front of a warm fire inside a tipi, my Blackfoot Native American friend beside me. After her greeting, she asks me to go outside to observe the symbols painted on our covering. I follow her direction.

Upon returning, I express to her my delight at seeing otters on our tipi. She explains that, according to Blackfoot legends, an animal appears to a human receiver through a dream or vision. The animal tells the receiver how to

paint his lodge and how to make other objects related to
it. Upon completing the tasks, various rituals are performed
in relation to the animal and the animal bestows some of
its powers upon the receiver. Since not every member of
the Tribe is so honored, these painted tipis are very sacred.
Of the smaller depicted animals, the otter is most prevalent.

I open my eyes.

Although I'm sitting alone, warmth spreads throughout
my being. I grasp my medicine bag and reflect, looking
upward to the spirit world.

"Thanks, dear friend, for sharing your ancestors' bond
with otters."

I rise from Mother Nature's embrace and slowly
assemble my gear back into its proper placement. My
shiny black snowsuit creates a fine contrast against the
whitened world.

I almost look like an otter!

The glossy outer fur of our otter family is dark chocolate brown on the upper surface of their body, blending into a lighter brown underneath; but when they emerge from the water, they look almost black. I even have their whitish muzzle and throat. (Mine isn't permanent; it's just caked with snow!)

Since we look so similar, should I attempt some of the otters' winter antics?

* * *

Before the snows, I bent over a slope to take measurements of a fresh otter slide and tracks. I stretched too far and inadvertently slid on my belly, just like an otter, down a slick mud bank into the gooey muddy waters.

"Is this supposed to be fun?" I yelled out. Some choice expressions followed.

There was no response to my question, except my own mutterings. While trying to rise to my vertical position, I continued to slip and slide, unable to acquire my footing. Finally, with help from Mother Nature's extended hand, I stood. Dirty and somewhat frustrated, I now had some understanding of how an otter slides.

* * *

Our otters slide on snow and ice during winter, and on slick mud banks during spring and summer—just as I did. In some cases, sliding is an easier form of locomotion than walking or running, and perhaps more fun. Otters may also consider it a form of amusement.

In a book published back in 1909, E.T. Seton wrote in a natural history account of otters and their slides: ". . . this is the only case I know of among American quadrupeds where the entire race, young and old, unite to keep up an institution that is not connected in any way with the instincts of feeding, fighting, or multiplying, but is simply maintained as an amusement."

I agree. So now, let's watch our otters slide.

* * *

During the night, Nature adds a thicker blanket of white across the Rockies, creating an environment for even more otter fun, and enabling the otters to reveal their sliding strategy.

They get a good running start from the top of the hill, then lie down on their bellies. With their front feet held along their sides and their hind feet out behind, they become a streamlined express. And the winner is . . . Mother Otter!

Well, after all, she is larger, has more downhill inertia, and much more experience.

The youngsters want to try it again. Back to the top of the hill and down again. The more often they slide, the slicker it becomes, until that great white puff in the sky releases more soft powder.

As winter progresses, the otters encounter new challenges in their search for food and den sites. When open waters are not available to them, our air-breathing mammals must search for food below the ice. That sounds like quite a challenge. However, the combination of Mother

Otter's experience and Nature's helping hand allows the family to meet their challenge.

Continuing cold weather reduces runoff of snowmelt into the river, thus allowing water levels to drop below the surface ice. Shelves form between the surface ice and waters below, providing good air-spaced corridors for our otters to travel and forage under the frozen cover. Other air pockets form along the outer arcs of curvatures in the riverbanks, and small breathing holes extend periodically through ice along the river.

Dens their mother chooses for the family are above water level, but they have openings under the ice for entrance into the water. Her knowledge of travel corridors and lodgings is necessary for the family's livelihood and for teaching her offspring survival techniques.

Mother Otter often guides the family to beaver-impounded streams and ponds. Here, they have access to their benefactors' lodges and bank dens, as well as to deep-water fishing.

Then, a familiar site appears on the horizon!

The family returns to the lodge where the young first began to learn their fishing skills. This special place seems to bring a smile to the faces of the little otters. Although the pond is now iced over, the lodge is empty and they know about the underwater entrances into the prey-rich environment.

I can observe the otters' behaviors above surface, but not while they're submerged. However, the otter's spirit shares their underwater adventures with me.

The otters have access to good food sources in locations to which their experienced mother guides them, and they have good lodgings that allow them access to air and water. I observe their streamlined shapes moving swiftly and gracefully through waters below the blanket of ice.

It is becoming difficult to tell the difference between mother and offspring. The young ones are growing up.

But will they remember the many lessons they were taught during the seasons? Will their learned behavior and instincts bring them through the years ahead?

Time is drawing closer when they will find that out for themselves.

The snow grows deeper and deeper as the weeks pass. Although the otters mostly slide on the crest of the snow, they sometimes tunnel through layers of down in order to reach their denning and foraging areas. They resemble small torpedos as they work through the mass of white.

What will they do if they meet another otter tunneling from the opposite direction?

* * *

I try not to break through the snow and crash any of their potential otter passageways. Most of the time I remain on the upper crust; on occasion, I break through. Then I topple over into the snow or take my own slide down a snow-clad hill, really trying to maintain a sense of humor.

During one such slide, when reaching the bottom of the hill, I just roll around and play in the snow. Even a dedicated researcher must have some fun. Before trying to rise to my vertical position, I lay on my back and gaze up into the deep blue sky, watching the white cotton-candy clouds sculpturing themselves into surreal shapes.

I see our spirits at play. My imagination allows the real me to join them.

Does this also occur to the otter male?

While in this higher dimension, I look down onto the Earth below. From one perspective, I see the otters at play; from another, I take in the accumulated ravages of humans upon nature. Tears of sorrow flood from my being as I watch Mother Nature's world ravaged by those who selfishly gain by its destruction.

I don't want to return to my real self.

What can I do? Will the battles for those of us who care ever stop the greedy among our species?

My imagination brings me back to Earth. I grasp onto Nature's body for assurance. She holds out her hand to help me rise. Although the ambient temperature is below freezing, I feel warm.

Yes, I will try.

My trekking takes me to many different types of habitat. One such location is the meadow where, during autumn, elk presented their rutting performance. The elk are silent now. But during one of my winter treks, another composition springs forth: howls of the coyote.

> *Upon this walk, I heard prevail*
> *coyote howls from nearby trail.*
> *The snow prevents my quickened pace,*
> *so I took Nature's soft embrace.*
>
> *Sounds from the wild moved through my head,*
> *but I really did not dread.*
> *They are part of Nature's tree;*
> *therefore, they must always be.*

Surprisingly, I see coyotes on only rare occasion; but I do encounter their signs. While walking toward the river on this particular day, all at once a crescendo of howls come seemingly out of nowhere. I look around but see no animals.

Are they echoes from the past?

I feel like I am standing within a sound system that surpasses any of man's making. Howls come from all directions. Standing quietly, I listen to this ancient symphony. I want to join in on the howling frenzy, but sensibility tells me that might not be a good idea. Snows hold me captive. Although not fearful, I do feel concern.

Then, as suddenly as it began, it ends. Still, I see no animals. I feel certain that the howls are real and not just in my imagination.

But where are the animals? Are they voices from the Ancient Ones?

Grateful for the experience, I decide this portion of river can wait until another day. I return to my vehicle.

Coyotes can be a problem for the otters, especially when the otters are small. After all, the coyote is a predator, as well as a cunning trickster. If a coyote were to encounter an otter with a nice large fish on land or an ice-covered waterway, it will gratefully accept the otter's gift. Although such encounters do occur, I never observe one nor does the otter's spirit share such an experience with me.

* * *

Not only do the young otters learn from their many lessons, they have fun along the way. They seem to particularly enjoy their sliding adventures. Sometimes, they slide across the ice of the waterways; other times, they slide down slopes of varied height.

One day, the family winds their way to the top of a relatively high hill. The young ones look over the side.

Is this going to be another lesson?

Before they have a chance to find out, a moving snowman appears before them. They watch as it begins to walk down the steep bank.

Wow! What is that?

They are not seeing a snowman, but a mountain goat—an animal the color of winter.

The surefooted mountain goat can easily move up and down sheer cliffs under varying weather and terrain conditions. Their hooves are well adapted to their lifestyle. Although this animal is not a predator, the otters need to stay clear of those dagger-like horns protruding from its head.

Mother Otter doesn't appear too concerned about the mountain goat and vice versa. The young otters, however, are curious about their new encounter. Maybe it will play with them. Then, a smaller version comes into view.

The young otters move towards it. Each species seems curious about the other as their senses work to evaluate their respective new discoveries. The kid butts its head toward the young otters. They quickly retreat.

I position myself a distance away from the two species and observe their antics through my binoculars.

The young goat was born two months later than the otters, but it is already at least twice their weight. Fortunately, the lethal design of the kid's horns won't be a problem during an unlikely physical encounter because they are only about two inches long.

The three young animals maneuver around each other scenting and making playful gestures. No actual physical contact occurs between the two species, but each mother still keeps a watchful eye on her young as they engage in their respective new discoveries.

* * *

Interestingly, the mountain goat isn't a true goat. It belongs to a small group or tribe of animals, sometimes referred to as goat-antelopes or "goatalopes." There are at least four genera in this tribe that are bound together by specially designed horns and thin-boned skulls. In turn, these anatomical features develop to suit their behaviors and the evolutionary niche they occupy in their mountain world.

I studied members of two of these genera—goral and serow—that reside in high mountain ranges in portions of Asia. Thinking back on those years of research, I notice a similar resemblance between the young mountain goat and the goral.

Although I conducted my studies in a large atypical captive environment, I saw similarities between the behavior of my animals and what is known of their wild relatives. My visit to the natural home of the Japanese serow in northern Honshu Island offered me a special experience to observe them in the wild. This animal is a visually appealing species that resides in rugged mountains of the three main islands in Japan. Their winter fur is an artistic blend of black, gray, and white with a fluffy mane encasing their faces. They are a solitary, secretive, shy species, so I feel fortunate to have observed several individuals in the wild.

* * *

After several minutes, the young otters and kid seem to lose interest in each other, so Mother Otter decides it's time for their departure. The threesome get a good running start from top of the hill, then lie down on their bellies for the longest slide of the young ones' lives.

What fun for these sliding cavaliers!

While getting caught up in this joyous occasion, I watch as the otters slide down the slope. Even the mountain goats stand and look in their direction. The kid jumps up in the air. Perhaps it wants to slide too. (But realistically, it shouldn't try.)

"Oh, to join them would be so much fun!" I sing out.

Through my imagination, I try sliding alongside the family. I weigh more and must use my appendages to thrust myself forward, so my progress is slow. When the otters' inertia wanes and they must give themselves a push, it is far smoother than mine.

Oh, well.

I return to reality, replace my gear, then begin my descent while grabbing onto extended arms of trees along my path. I reach the bottom of the hill, but the otters have long since departed from the area. However, they left behind clues for me to follow.

I try to catch up with them, but Mother Nature places obstacles in my path. Fluffy aerial sculptures begin to release their white downy feathers onto the land. As I follow the otters' trail, the snow intensity increases while the ball of light that has brightened the sky starts its descent behind the mountain backdrop. The combination of heavy snow and lack of light will soon envelop me.

The otters continue onward to a river-bank den. From here, they enter the water for a much needed meal. Our sliding travelers are both hungry and tired. After fishing and devouring their meals, they prepare for a good night's rest.

More adventures lie ahead for the otters, but the young ones will soon experience them without their mother. The end of their first year of life is drawing closer. A major change will soon occur.

But for now, I visualize the young ones sleeping and dreaming of the happy times they had with their mother, dreaming and remembering the lessons she taught them along the way.

My hope for their future is that the great Otter Spirit will always be there to guide them on their path of life.

CHAPTER EIGHT

Dispersal

Many lessons the otters were taught;
now to apply them with serious thought.
Although they'll still have some time to play,
soon they must begin to dawn a new day . . .

Dispersal

*C*old breezes sweeping across the snow-clad landscape sculpt asymmetric shapes into the terrain. As I watch, my imagination turns them into visions of familiarity.

Whispering winds blowing through needles of pines create dancing shadows on this sunlit stage. Nature's ballet begins its performance with a graceful adagio, seamlessly glides into andantino, then whips into an allegro. Shadows begin a lively dance as the crescendo of wind howls through pines. The exuberant movements would soon exhaust a professional human dancer, but Nature's ballet continues.

I pull further into my winter-clad covering.

"Maybe I should join them just to keep warm," I contemplate in my trembling state.

Are the otters also watching the performance or are they snuggled inside their nice cozy den?

I am eager to determine that answer.

Eventually, the winds return to a pianissimo and the dancers to an adagio.

I stand to make sure I haven't become a part of this frozen landscape. I'm cold, but feel fortunate to have witnessed this special event.

Throughout this long season of the year, both the otter and I continue our respective trekking and learning ventures. Sometimes, I directly encounter the otters; other times,

I play detective, following clues they leave behind. However, the two little spirits never stop teaching our opposite species' Earthly forms.

* * *

The calendar shows that spring arrives soon. This means much more to the young otters than a new season of the year.

Do they sense what is coming? Do changes in their mother's behavior towards them send a message? Does the onset of longer days convey a piece of information? Or is there just some unknown factor that creates restlessness and a desire for independence within them?

River otters have a variable dispersal strategy. There isn't one answer that applies to them all. We do know, however, that a big world awaits our young ones.

Will they be up to the challenge? Will I be up to the challenge of following them?

Determined that the youngsters must establish their independence from her, Mother Otter spends more time separated from her offspring. She doesn't forcibly chase them away when they seek her company, but she imparts hints and clues. When she catches a fish, for example, it is for her meal and not theirs. Nipping at them on such occasions creates confusion in the young ones, but they soon realize there will be no more free lunches, and they begin to forage successfully for themselves.

Brother and sister remain together while exploring within areas of their schooling during the transition to this new paradigm. After all, they have never been separated from each other or, until now, from their mother.

Conflict ensues until one day an unknown force sends the siblings running, sliding, and bounding across the ice and snow to begin their journey toward independence.

* * *

The foreshadowing of this event seems somehow unfulfilled. It just happened without notice or fanfare (not that they should have served formal notice on me that they were leaving, mind you).

Suddenly, however, the youngsters are on their own.

I sense fear for the young otters, particularly for my special one, since our bonding continues to intensify. I desire to protect him on life's journey as did Mother Otter during his nascent year. Although I will always try to be there for him, that is not realistic.

I call on the great Otter Spirit to be his guiding force.

Gliding and sliding over ice and snow, I follow the young otters. On my trek, I approach an area of beaver tracks and signs of branches being dragged across the snow. It is now the end of March, so beaver activity has begun for the new season.

I sit next to a lodge. From within, a gruff sound emerges.

Remembering those sounds from my previous encounter, I softly apologize for the intrusion and walk away. While looking back at the beaver marks on the snow, I giggle while recalling another beaver-related event that took place during the past summer:

One of my trekking areas along the river
extends out from a campground. I see marks on the

bank, leading into the water, that appear as if someone raked the shore. I think that rather strange and couldn't imagine anyone spending time to rake this natural area. Finally, one day I find a broken tree branch on the bank and pull it across an unmarked area of shore.

"Ah-ha, my dear Watson! The evidence points to none other than a beaver as the raker, dragging branches into the water," exclaims Ms. Holmes.

Nature provides us with so many interesting mysteries to solve.

I lost the otters' signs I was following. My last encountered clues were slides leading through an ice hole into the waters. Now, trying to find the otters in this beaver pond and channel system seems almost impossible. I'm not sure which way to go.

Then, two Canada geese land a short distance away.

Ah, harbingers of spring! Favorite birds of my Native American friend.

"Yes, she extends her thoughts to me from the spirit world," I muse while sitting on a bank of snow. I grasp my medicine bag for guidance.

Upon again rising, I feel drawn to a small area of open water. On the bank, I find two clear slides where the otters pulled themselves up from the waters. Beyond the slides, I spot tracks leading in the direction of their travel.

"Thanks," I say to the geese as I look up to the sky.

In my joyful, playful state, I feel like an otter. Many North American tribes attribute such qualities to the otter.

It is a species that brings together the important elements of Earth and water because it is at home in both. In a tale I once read, the comedian Otter is caught by the enemy and threatened with drowning:

"Oh no, not that!" Otter screams out as he is tossed into his own element to laughingly swim away.

Ah, yes, the happy little clown of the animal world brings a smile to the lips of many.

Trudge, trudge, trudge—I reflect on Otter's special attributes, and my frustrations make a 180-degree turn. Trudge, trudge, trudge—I keep pulling myself through the snow . . . with a smile.

To those of us residing within this montane community, it seems like winter will never end. However, both the otters and I persist.

Do the otters ever tire of their increased struggles during this time of year? Even though they are well adapted to and playful within the snows, ice-covered waters, and cold temperatures, do they secretly look forward to a break from Old Man Winter?

Sometimes, I wonder if the slow emerging spring will ever return to this mountain terrain as Mother Nature's crystal tears continue to fall upon the land.

I sit down beneath a tree.

Huddled within Mother Nature's arms, I recall a legend I once read from the Native American Senecas of the

Northeast woodlands, entitled "Spring Defeats Winter."
The title is certainly reassuring. Briefly told:

> *When the world was new, an Old Man with*
> *long white hair moved throughout the land. His*
> *presence caused land to become hard, waters to turn*
> *solid, plants to die, leaves to fall from trees, and*
> *fauna to flee. He and his only friend, North Wind,*
> *would talk and laugh at the things they did to make*
> *the world a cold, difficult place.*

Hearing their laughter, I draw deeper into my winter-
clad clothing. My imagination carries me into Old Man's
lodge on the day Young Man arrives at his door:

> *North Wind sensed a change occurring, so he*
> *fled. Well, this uninvited guest entered the lodge and*
> *began stirring a fire that had previously given no*
> *heat. Warmth spread into the shelter making Old*
> *Man angry as his cold world began to shrivel. Old*
> *Man insisted that he was the powerful one, causing*
> *strong Young Man to laugh. Then, Old Man melted*
> *away as Young Man's companion, South Wind,*
> *blew her warm breath into the environment. Yes,*
> *Young Man Spring defeated Old Man Winter.*

I stand up to continue my trekking. "It shouldn't be
too much longer."

* * *

Spring soon arrives, and none too soon for those of us not naturally adapted to long, harsh winters. With spring will come the foretold changes in our otters' lives.

During the young otters' time with their mother, she moved the family through several sections of land and waterways that comprise her "home range." This is a chosen area in which an otter lives, reproduces, and satisfies its life requirements. Although variable, the land configuration must be connected to water.

The size of an otter's home range also varies, depending on the region of the country in which it occurs. In our mountainous location, otters' home ranges can include land and waterways encompassing 8 to 38 linear miles. Habitat variations along the course of the river affect the size of the home range, as does an otter's sex, age, season of the year, and related food supply.

Winter home ranges are smaller than summer ones; those of a female with young dependents are smaller than those of a single adult otter. Our mother otter may possibly share her home range with more than one other otter, particularly a male. Male otters' home ranges often overlap with more than one female, as this is advantageous during breeding season.

As the yearling otters explore and move throughout various watersheds, they will learn the signs indicating that other animals occupy an area. The young male, in particular, will learn very valuable male lessons. But for now, our young explorers continue moving through spring and discovering their world of independence.

The siblings swim together in the river, sometimes stopping off at known beaver-constructed channels and ponds. Their fishing excursions are not always successful, but they learn to supplement their diets with other aquatic prey such as insects, small reptiles, amphibians, and freshwater mussels. They must keep their tummies as full as possible to sustain their active energetic lifestyle. Though they try, they can't always eat the nearly two pounds of food required for their daily diet. Sometimes, they go to bed hungry. But Mother Nature's warm smile will soon melt her ice-covered waterways, providing the otters with an increase in prey selection and varieties of fish that are easier to catch.

The behavior of our two yearlings indicates they savor spending time together during this sometimes traumatic period of their lives. They often stop to play and even groom each other. One day, while frolicking in the nearby forest, an adult coyote suddenly emerges from a thicket of trees and dashes towards them.

Frightened, the otters run, but in different directions. The coyote follows the smaller female. A cry of alarm resounds through the forest. The male stops, turns. Recognizing his sister's call, he runs towards her cry for help.

I am trekking nearby when I hear the commotion. Trembling, I yell loudly while running in the direction of the female's scream.

Suddenly, all is silent!

My entire body shakes uncontrollably as I investigate the area of disturbance.

Where are they?

Two heads, slowly, peek out from the roots of a large tree.

I immediately recognize the distinguishing streak above the male's eye, telling me the youngsters are safe. Standing before them, I bury my head in my hands as tears of relief steadily flow.

The youngsters turn and dash back to the river. I turn and walk back into the forest. I'll never know whether my intervention made a difference.

Hopefully, they each learned an important lesson in survival, even though coyotes typically don't prey on otters. This was likely just a female coyote protecting its young. But had our otters been fishing, the coyote would not have left empty-handed—one way or the other.

The yearlings continue onward in their life's journey. While swimming in their world of increasingly open river waters, they meet two other otters. Through their otter language, they learn that each pair consists of newly independent siblings. Although the social structure of otters is fluid, coming so close to other individuals is a new experience for our pair.

When very young, they saw another otter in the distance; but their mother's behavior at that time indicated they should avoid the other otter. However, now on their own, our siblings join this other pair, spending several days swimming and playing together. They forage in the same general locations, but each otter usually does so independently. On occasion, they use cooperative fishing efforts.

The four otters appear to enjoy each other's company. For the time they are together, maybe they feel a security in numbers against the predators of the natural world. Sometimes, they even stop to engage in an "otter dance." (Remember the proposed "beaver dance?")

When the otters reach a segment of land that contains a form of their species' communiqué, they move onto shore. Their particular chosen location is where the resident otters defecate, urinate, and secrete from their anal glands. This designated location is periodically revisited by the resident otters. Some of these land segments are known as "activity centers," where otters of the home range have a particular site attachment. They let other otters moving through know that this area is taken. Their elimination and secretion behaviors are a form of marking.

Well, the youngsters figure the warning doesn't apply to them, at least not yet. So, one otter takes the lead and is the first to investigate the area. It performs dance steps.

©MR04

These steps include alternately treading their back feet while holding their tail high in the air, followed by elimination. This signal alerts the other three to follow suit.

Even otters dance in activity centers! We humans could perform the dance itself, excluding, of course, the elimination and tail-raising behaviors, and name it the "otter shuffle."

* * *

Spring soon gives way to summer and temperatures grow warmer, at least for the Rockies. The two otter pairs separate, each going their own way. Our siblings swim west along the river and farther from their mother's home range.

I try my best to keep up with the young nomadic otters. Sometimes, I do encounter the siblings. Realistically, however, finding the otters and following their signs in snow is much easier than during the summer.

At least, the young male's spirit keeps me apprised of their activities. It also guides me to their particular locations; but often by the time I reach the area, they have already moved onward.

I concurrently continue my otter survey, so my time is never wasted. When wildflowers start emerging, I work to determine their identity. Although fun in the beginning, as the warmer temperatures progress, new wildflowers pop up along my path on an almost daily occurrence.

"Oh well, I guess I'll just enjoy their beauty and let the botanists figure out their taxonomy," I laugh as I give up that impossible task.

One day, as I sit next to a small pond nicely carved out from the river's flow, shadows of leaves glide through the waters, reflecting shapes unknown to their masters. One lone, browned, ragged leaf, floating atop the surface, mirrors a perfect flower within the clear blue waters below. Other leaves, still fresh and new, do not reflect the beauty of this aged one. Then suddenly, dancing water striders, whose shadows appear as tiny monsters within the deep, enter and disrupt the tranquil scene.

Simplicity and yet complexity always exist in Mother Nature's garden.

* * *

Times become easier for our young siblings. Their foraging behavior increasingly improves. They find a variety of empty comfortable lodgings near the river. They explore land bridges between different waterways and beaver-created systems. Sometimes, they take time to play and chase each other, then stop to wrestle amongst the arrays of colorful wildflowers. All seems perfect in their world.

However, each one appears to sense that another change is about to transpire. Although they enjoy each other's company, they are being drawn in opposite directions. The young male's more independent nature begins to manifest itself as the mighty river keeps calling him to continue his westward journey. The young female's behavior betrays her desire to return eastward and back to more familiar territory. Maybe she will again find her mother, or be accepted by another mother with young offspring that can use a helper. Her innate need for duties that will eventually help her, when she someday bears her own young, are becoming evident.

One day, a small group of male otters appears in their world. They are moving in the same direction as our young male desires, so our male swims away with the group. The immature female turns around and swims in the opposite direction. As is natural for otter siblings, this is probably the last time the two will encounter each other.

* * *

The weeks of summer come and go very quickly. Cooler breezes of autumn begin to fill the air.

Although our yearling male remains with the group, he begins to sense it's time to leave them. After all, when winter approaches, food becomes less available and difficult to access. Competition can be fierce. His mother took care of that problem when he was little. When he and his sister began their independence, they stayed in an area of familiarity. Now, his sister is probably back there again.

What should he do? Should he also return? Or should he continue searching for his own space?

For now, at least, he decides on the latter.

Our independent sibling strikes out to seek his own domain in the vast watershed of Rio Colorado.

CHAPTER NINE

Strange Encounter

Large bones emerge from one of the past.
Enter a species for some contrast.
Although encounters may once have occurred,
none quite so strange as this now observed . . .

Strange Encounter

The young male otter continues to swim west in his aquatic environment. He is still searching for his own space in this riparian world. He finds too many "this land is taken" signs along his path. Certainly, there must be a place for him too.

He considers going back to more familiar territory, but for now he continues onward. He pulls himself out of the water and onto the bank for a rest. Then, he moves a short distance inland to search for a den site.

What is that which lies ahead?

It looks like a large tree, toppled onto the forest floor to begin its cycle of decay back into the Earth. The otter has used many similar sites for his rest, so he moves toward the object.

Yes, it looks like a good place to lie upon to get sun and underneath to get rest.

Up to the top, he progresses.

Hmmm, there are interesting crevices to explore.

The relatively smooth surface of this "tree" brings forth a new sensation unlike that of other trees encountered in his short life span. He tries to bite on parts of its protrusions, but it has a very hard surface.

Well, he'll figure that out later.

For now, he lies down for his sun bath. But soon, the large yellow ball begins its downward descent. Since the otter ate during his daytime adventures, he is already full and ready for sleep. Thanks to the diggings of another animal, he finds a place carved out underneath the object of his discovery. He slides into it for a much-needed rest.

The otter closes his eyes.

My little spirit enters his dream world.

Beyond, he sees a world unlike any he has ever seen before. There are expansive flora and flowing rivers, and exceptionally large, strange-looking creatures moving around in this world.

* * *

Actually, that smooth "tree" the otter discovered and underneath which he lays is an uncovered bone of a mammoth, an extinct contemporary of the current-day elephant. The ancestors of this species, and of other extinct and living elephant forms, date back in the geologic time scale to the Eocene epoch—55 million years ago—on the continent of Africa. The origins of the otter's taxonomical animal family *Mustelidae* are thought to date back to the Oligocene epoch—38 million years ago—in Eurasia. This family contains about 25 related similar genera. The lineage may actually date back that far in time. But paleontological evidence of the time and continent of origin of this disparate family continues to emerge.

Our male otter will likely encounter members of his extended family—badgers, ferrets, martens, minks, skunks, and weasels—in his lifetime; but there are others that live in different parts of North America and on other continents that he will never meet.

Paleontologists have found evidence that the mammoth evolved from its lineage approximately three to four million years ago. Evidence for our river otters indicates they evolved from their linage about two million years ago, although animals with otter-like characteristics show up several million years before that. The fossils of one species found in North America date back perhaps as many as six million years. Some speculation suggests that this species evolved into a relative of our otter now found only in South America.

It is believed that mammoths entered the New World during the Pleistocene epoch (about one million plus years ago), sometimes referred to as the Ice Age. These animals, along with many other migrants, including the ancestor of our otters, moved back and forth between the Old World of Eurasia and the New World of North America during this sustained period of time. They used the Bering land bridge, a connecting land mass between current-day Siberia and Alaska. This bridge was created during a glacial period when the sea level dropped low enough to expose the connecting isthmus.

Our otter's current movements across "land bridges" between waterways along the river are similar, but on a much smaller scale than those undertaken by his ancestors.

A number of ancient animal forms emerged from what is now Alaska and Canada; they then radiated throughout the continent. Another species of the elephant family, the smaller mastodon, actually arrived in North America during a different glacial period, the Miocene, about 16 million years before the mammoth. Even some members of the otters' family may have arrived during this same geological period. Both the mammoth and the mastodon became extinct 10,000 to 11,000 years ago. Although elephants no longer live in North America, otters and other members of its taxonomic family certainly do.

Researchers continue working to figure out more of river otters' ancestral movements. For smaller animals, much of the evidence is based on individual teeth and mere fragments of bones recovered by paleontologists. So far, discoveries of fossils from our otter's species in caves in North America date back from 750,000 to one million years ago.

* * *

The young otter moves around in a restless pattern. The spirit never before showed him anything so large and shared with him anything so confusing.

He awakens and decides he suddenly feels hungry. He peeks out from underneath the large bone. His physical world hasn't changed. He doesn't see any large animals moving about like those of his dream.

Swiftly, but cautiously, he slides into the nearby river for a snack. Maybe he should just keep moving. Yet, something tells him to return to his now-familiar resting place. He swims upstream a short distance and then back to get some exercise and fill his tummy. He walks back to that unusual "tree," moves on top to once again feel its surface, then goes back underneath for a rest.

My joyful little spirit re-enters the dream world of the male otter, this time to share with him something he will not physically encounter.

Even though the mammoth, under whose bone he sleeps, will never again naturally enter the world, there evolved two current-day genera, African and Asian elephants, that live on their respective continents. In the course of my career, while spending several years studying African elephants, I discovered some of their secrets just like I am discovering some of the otter's.

Through the otter's dream world, he and I travel together to the open savannas of the African bush elephant—a very different habitat from that of his own. We watch a group of elephants drinking and bathing in a waterway, and decide to join the young ones in water-related play. Splashing water towards us with their trunks, we respond with squeals of joy. Although this species is terrestrial rather than semi-aquatic, they do enjoy their time in the water.

The elephants leave the water, dust their bodies with soil, and walk from the area. The matriarch vocalizes a low-frequency rolling sound to call her family together, sending vibrations through the otter's body during his dream. Eventually, he relaxes and we follow the elephants in their search for forage. At least, we feel comfortable in the fact that they are not carnivores, but instead feed on grasses and other vegetation similar to the deer species of our own country.

Our spirits remain with the elephants for a while to observe them feeding. Because the adults can weigh 500 or more times the weight of our otters—and have a

relatively inefficient digestive system to boot—they must spend most of their day searching for food.

We spot a group of frolicking youngsters. They engage in play behaviors similar to those of our otter and his sister. They, too, wrestle, roll and tumble, chase each other, and manipulate natural objects. However, they don't engage in sliding, over and over again, down an embankment of mud or snow, at least not on purpose. But just imagine, they would look kind of cute sliding down a hill.

Perhaps the otter thinks about the times he played with his own sister. Joining the young elephants may be fun; but realistically, the otter may sense that they are too large to play with.

Before we return to our own continent, we decide to search for some individuals from other otter relatives. Two species of clawless otter and the African spotted-necked otter are both endemic to Africa; the Eurasian otter also has a population here.

Ah, what fun it would be to swim and play with one of them.

Our spirits follow a river course.

Up ahead we find an otter. We join him, and the three of us dive and swim within the warm waters. Sliding onto shore, we chase and play upon the land. The special otter and I enjoy the antics with our new-found friend. In fact, we almost forget where we are until we spot a group of elephants.

Since this African otter species encounters elephants during its wanderings, was this also the case for the extinct elephant forms and the ancient otter species in North America?

Something for us to contemplate as our spirits return home.

The otter awakens and finds himself alone in the den underneath the mammoth bone. Confused by what he did or didn't experience, he becomes restless.

Eventually, he returns to sleep.

* * *

I hope to once again catch up with the young male.

At this point in his wanderings, it becomes almost impossible to reach him by walking the river. He seems to be always on the move and one step ahead of me. I drive as close as possible to the place the otter's spirit guides me, then trek to the watercourse in pursuit of my special one.

This time, I follow directions to the bone, both figuratively and literally.

What an amazing discovery!

I slide my hands across the bone's surface. The sensation provides me with the imagery of that long-ago world.

Wow! Now, if the otter is only still resting underneath his discovery . . .

As if on cue, the young male slides out from his night's lodging. The unique identifying white streak above his eyes seems so prominent now. He looks around to make sure his world hasn't changed.

No, it still looks the same.

But before he heads to the water, he turns and sees me. Our eyes, our hearts, and our souls meet for that one brief, that very special moment.

Does the otter feel the same way as do I?

"Please don't be afraid," I plead.

I sense he understands.

"Yes, special one, we are two different beings on a distant shore of life. One day our spirits will coalesce into One. For now, you are very young and still have many more adventures in your journey through life. Even if I can't always physically find you, my spirit and yours will continue to teach and guide our Earthly forms. Please go in peace," I urge, as I motion towards the river.

The otter turns and slowly moves to the water, then slides into his milieu.

He swims around making playful gestures. His movements seem to indicate indecisiveness as to whether to go or stay. However, after that strange night, he is very hungry. So once again, he heads westward in search of nourishment and his new life.

I try to follow the male, but his rapid movements make it impossible for me to catch up.

I return to the bone to take measurements and photos. This bone means so much more to me than just being from an extinct life form. It is a connection to both the first and last animal of my research career.

As I rest my body beside it, my thoughts enter into an ethereal dimension.

Soon, I fall asleep.

CHAPTER TEN

The Quest

West I swim, though times get rough;
but I'll prove that I am tough.
One day, I will find my space;
therefore, I'll keep up the pace . . .

The Quest

Another winter approaches the Rocky Mountains. Aspens and oaks drop their colorful leaves as Mother Nature's breath brings a chill to the air. The young male otter will be soon put to a very difficult test.

Will he remember all the lessons his mother taught him?

He and his sister had some difficult times when they first struck out on their own. Snow and ice still covered much of their world. This time, however, he must endure an entire winter season, quite possibly strictly on his own.

Would life have been easier for him if he lived during the time of his ancient ancestors and those huge elephant forms?

The season's first snow begins to fall gently upon the Earth. The otter pulls himself up from the water onto the shore. He lies on his back and plays with the soft flakes falling onto his paws. He watches the smaller animals scurry around, busily increasing their food stashes in preparation for their own winter survival.

Maybe he can build a stash of fish. All he needs is a freezer!

Well . . . it was a thought, but live moving prey is a part of his innate heredity.

As darkness slowly enters his world, he returns to the river for supper before seeking out a place to rest. When his tummy is full, he moves back onto land to search for a vacant room. He soon finds a nice den created by a fox that isn't staying there at the time. He moves inside for a good night's rest.

The otter awakens early the next morning and peeks out from his den entrance. He still feels uncomfortable about the world shown to him by my little spirit. He wants to make sure nothing has changed.

Instead of seeing large animals moving across the terrain, he is greeted by a blanket of white, covering the ground upon which he tread the night before. Winter has returned to the Rocky Mountains. Although the young one hoped the season might enter more gradually, he can do nothing about it.

He slides across the carpet of white and into the cold waters. At least, the river didn't freeze over during the night.

* * *

Moving around in winter is physically draining on me. However, it is the time of year when I find more clues about the otters' movements, along with those of other animals. I think back on a Nature story revealed to me last winter:

> *While doing my otter tracking, I come upon clues left in the snow by a predatory event. Following the tracks of the predator and tunneling movements of the prey, I reach the place of their encounter.*
>
> *"Ah-ha, my dear Watson, these clues tell me that the predator was a coyote and the prey a tiny vole," sings out Ms. Holmes.*

Yes, another case solved.

Mother Nature again covers her body, her world, with soft flowing robes of white, while her gentle voice stirs winds of change. Taking a break from my tracking to sit on her soft sheet of satin while embraced within her arms, I close my eyes and listen.

Sometimes I hear total silence, sometimes the subdued sounds from a bird or a mammal engaging in its respective pursuit. At still other times, I hear the flute-like sonance from open waters as they flow through a passage in Nature's ancient symphony.

Absorbing this silence, these sounds, into my soul, I continue my work refreshed, floating through space and time. I am invigorated with the realization of having become perceptive of something beyond most humans' conception . . . because I listened.

The soft white flakes tease me in this often playful time of year. I hold out my hand, like the otter holds out his paw, trying to grasp tiny speckles of white. Each of us has a similar experience.

Each of us is connected through the worldly guidance of the great Otter Spirit.

* * *

Seasons, and then years, float the otter and me through tides of change—almost as if we were accelerated through time. Five years elapse almost as if it were one. Each of us has many experiences and is taught valuable lessons. Each of us continues our respective quests in life.

During the otter's travels, he discovers that the perfect world of his youth is not always projected onto everyday life.

One day, while swimming through one of the river's extensions, his sense of smell detects an unknown, pungent odor. He pulls himself upon the shore and watches as fish float by, atop the noxious waters. For the first time in his life, he feels death entering his world.

My spirit finds the sad and confused otter and tries to explain to him what he sees:

"Either unreclaimed waste products from mining operations or illegal dumping of toxic materials has poisoned the water for all life that it sustains," I share.

"Some humans have little or no regard for Nature when it doesn't fit into their agenda. They foul her body and they foul her waters in pursuit of their own greed. In so doing, they destroy Mother Nature and leave behind a poorer place for their own future generations to sustain their lives," my spirit conveys to the otter. "All life is interconnected."

Then, my tiny white shape guides him away from the contaminated area.

Because otters are known to be a high-level indicator species (meaning they are most susceptible and at highest risk to adverse changes in their environment), our male learns a most important, if sad, lesson.

My spirit always seeks to guide my special one through both good times and bad. Feeling his confusion and sadness, I embrace him from afar with my heart and soul.

Finally, he reaches his quest. He finds his space—a section of waterway where there is no longer a male in residence. He finds his home.

Now he will be the dominant male otter in this space. During the mating season, he can court female otters and begin his own families. He endured the many challenges during the five years spent seeking his destiny. He pursued and he won the battle.

I, too, pursue my quest of learning more about my special otter's behavior. Though his spirit tries to guide me to his whereabouts, his wanderings often elude me. The many challenges he presents me begin to take their toll.

Since otters are very mobile, they usually will not stay in a small expanse of land-water area for a prolonged period. Therefore, the otter male and I are constantly on the move.

Finally, one day, I return to my former study site. Here I will continue my survey and try to find other otters, or even the male's mother or sister to study. I feel discouraged, however, because I want to continue with my special one. Fortunately, this disheartened state does not last.

To my delight, the male otter also returns to my former study site—in an area adjacent to the home of his birth and nascent year.

In his quest for his own space, the otter worked his way back to a location familiar from his youth. Perhaps the Otter Spirit guided him, since such a return usually does not occur in his species. Although not directly on the river system, this adjacent watershed includes creeks running into and out of two serene lakes. These are clean, clear waters fed by Nature's winter bounties and her tears of joy. This affords a good population of prey for his foraging activities. Unpolluted waters and a good prey base are most important survival requirements, as he discovered during his travels.

If Nature's life-giving resource is poisoned, the otter and other water-dependent species cannot live here. Fortunately, that is not the case in his new home. The land contains trees and other vegetative cover for his movements between the lakes and their tributaries, and for other land activities.

In fact, this beautiful environment is a favorite place of mine. Throughout the years, when not on the river system, I spent many hours in this location.

* * *

The male otter and I do not know, at first, that we both returned "home."

I again become actively involved in my survey work, which leads me into some rather unique circumstances:

One summer day, while trudging through tall grasses and climbing over fallen trees, I reach a system of holes within the sprawling roots of a large tree next to a channel of water. Tired, so not thinking clearly, I stick my head into one of the holes.

"Splash!" emerges from within the dark world below.

I quickly pull out my head and look over into the channel. Whatever was there is there no longer. (Imagine an animal seeing this blob of human flesh suddenly appearing in its nice quiet world.)

"I would scare me too!" I laugh out.

Signs in the area tell me the animal is either a beaver, an otter, or perhaps a muskrat. But I will never know for sure. One thing I do know: it is one frightened animal.

When my common sense once again takes over, I can't believe what I did. It's a good thing for me the animal didn't attack.

Another time, I lost, what to me is, a valuable possession. I often carry a small tape recorder on which I record various otter vocalizations, when fortunate enough to catch them in time. On occasion, I play these sounds, hoping to attract an otter. Because of my busy schedule, I never made a second copy of the tape.

One day, while walking out from the forest into a meadow of tall grasses, I fall into a hole. Not realizing I lost something, I pick myself up and continue my trekking. After walking a distance away from the area, I look for my tape recorder. It's missing! Assuming it dropped when I fell, I return to the meadow. However, the hole is so well hidden in the dense grasses, I still can't find it. Searching and searching is to no avail.

Before giving up, I grasp onto my small medicine bag and ask for guidance.

I walk to where I am directed and find my recorder!

Feeling very humble, I sit next to the hole and look up into the sky. I express my thanks as a magnificent osprey flies above my head.

I rise and walk back to the river. So absorbed in the moment, I don't realize the sun is finding its way behind the mountain, with darkness close behind.

However, my special guidance continues.

Circling above me are three red-tailed hawks directing me to the path I am to follow. Just as I reach my vehicle, the osprey reappears. I sense an important message being conveyed.

I listen with my ears and also with my heart.

"You and the male otter will again encounter each other."

* * *

The adult male periodically moves throughout his home range to mark specific locations as a sign, especially to other male otters, that there is now a dominant animal in residence. On these jaunts, he swims around the lakes or through his convoluted streams, stopping off at his designated shoreline locations of prominent boulders, logs, or openings in the vegetation, where he leaves his "calling card." Sometimes, he moves from one waterway to another across land bridges, which themselves often contain segments of wetlands for his communiqué. The most frequently visited sites are those which extend out from his favorite fishing holes.

Vegetation surrounding his aquatic areas includes species of riparian shrubs and plants that either require, or at least tolerate, a plentiful amount of water. These water-loving varieties are guarded by land-loving giants of pine. The forest floor of these tall ones is carpeted with areas of dense understory components. This provides various wildlife species with places to hide from predators, or even from each other. Often, little chipmunks engage in games of chase or hide-and-seek. If there is a winner, only they know. Whatever their game rules, it looks like they have fun.

During late spring and throughout the summer months, Mother Nature sprinkles lovely colors of wildflowers along our otter's path of travel. Sometimes, he encounters small umbrella-shaped "alien ships." He stops to bat them off their pedestals in his own devised otter game. They bounce like tiny sponge balls across the Earthen floor. Even adult otters invent solitary games to amuse themselves.

Our male may not be the only otter in this home-range designation. This aquatic habitat can easily contain more than one female, even with her recent offspring in tow. There really is enough habitat for everyone, including a good supply of food.

Early summer brings forth the spawning of one of the otters' favorites, a species of sucker, the slow-moving variety. Other resident fish species, fresh water mussels, aquatic insects, and a good supply of yummy crayfish add to his menu choices.

The otter's benefactor, the beaver, also inhabits sections of the lake and the large opening of one of the creeks. In addition to enhancing the habitat, this means comfortable lodges where the otter can rest. In fact, there are many places where our male can spend his nights and his days with very little disturbance. He has dens created by other species, along with logjams, hollowed logs, and boulder outcroppings.

The Otter Spirit guided him to a perfect otter environment.

* * *

One night in early September, the male otter's spirit enters my dream world.

He shows me the lakes and creek which are so special to me.

Can the real otter possibly be in that location? It is too much to hope for.

I rise long before dawn the next morning and drive several miles to the area. I bounce out of my truck and dance along the path, quickly reaching the lake. Clutching my medicine bag, I am guided to the pond at the large opening of its creek outlet.

I sit down to wait.

The sun begins slowly appearing above the surrounding mountains. In the smooth clear waters of the pond, Mother Nature mirrors her beautiful scene. Tops of towering pines sway gently in mild breezes while peeking down at their own reflected grandeur. Merganser ducks swim gracefully through the pond, stopping for a bath and frolic, their ripples abstracting the reflection until they subside. Two kingfishers fly across the water vocalizing and chasing each other before finally alighting to search for prey. An osprey calls out from the top of a tall pine.

I look up at the exquisite creature of the skies as I listen to his message.

Can it be the same osprey from the river that foretold events yet to come?

Ripples appear in the water and an otter appears!

The distinguishing mark on his head tells me at once that it is my special one. I know from not only my vision, but from my heart and soul.

I quietly thank the messenger; then the osprey flies away.

The otter glances in my direction. He senses I am not a threat, so he continues his activities.

He moves up onto a large boulder where he rolls on his back and grooms his fur, then rises and performs his elimination-marking behavior. Back he slides into the water to fish for a meal. He is under water for only a few seconds at a time. Success comes quickly. He moves back onto the same boulder and eats his breakfast. Then, he slides into the water again and swims to a nearby beaver lodge. After a few minutes, he emerges, swims back, and climbs up onto the boulder. He again grooms his fur, using the claws of his digits and the incisors of his teeth. Finally, he lies down for his morning siesta. He rests for about a half hour before moving from the boulder and into the nearby forest.

I can now confirm that, as did I, the otter returned home.

I rise and walk around the waters to the location where I saw my special one disappear into the woods. Deeper into the thick forest I trek. I come upon a mound of soil. Nicely imprinted in this barren patch of earth are mammal tracks. But they aren't from an otter. The animal that leaves this sign is much larger.

The tracks are from a black bear!

I look around for the big one, but he's not in my immediate view. Since *I* found this fresh track, I wonder if the otter did also. *Did the otter actually encounter the bear?*

The winds increase, causing the pines to whisper a message to me.

"I'm not sure what you are telling me, but I think I will return another day," I speak out loud to the towering giants.

I feel confident that the great Otter Spirit has protected my special one.

I walk as quickly as possible while constantly looking over my shoulder. When again reaching the waters, along the shore I find a small bunch of flowers. Next to the bouquet is a fresh bear track.

"What exactly is this message?" I ask.

There is no response to my question. But I decide not to stay and play "Ms. Holmes," even though my immediate interpretation is that of a peace offering.

Trembling, I quietly walk away.

Even though the majority of the bears' diet is vegetative, they are a potential predator of otters and intimidator to humans.

I reach my vehicle. Once inside, I sit quietly to regain my composure before driving to my own den.

It was a special day. But maybe I'm just getting too old for this type of work.

I look up onto the side of the mountain where strands of aspen cling to their green garments. Soon, their dressings will turn to bright yellow as they announce the new season.

I, too, am entering the autumn of my life.

I start my truck, then drive slowly toward my cabin. As my mind reflects back to the events of the day, I think about the complexity of life within Nature. Tears begin to flow from somewhere deep within my soul when I think of those humans who try to destroy her world.

Clouds shaped in the otter's and my Earthly forms begin moving through the sky.

I grasp my medicine bag.

I must continue my work.

CHAPTER ELEVEN

The Adult Experience

You've learned many lessons in your journey through life;
the teachings of mother have lessened the strife.
Now, as an adult, new challenges remain;
but when they're met, you will definitely gain . . .

The Adult Experience

During the male otter's first months in his new home range, he meets a couple of mature females. Although this occurs well in advance of spring mating season, at least he knows they are there. One of the females has two young offspring in tow, keeping this focused mother and teacher very busy.

One autumn day, our male watches her swimming gracefully through the waters with her young ones close behind. Grasping a large fish in her mouth, her offspring try to take it from her.

A lesson?

Their movements are very swift as they appear above then disappear below the surface of this calm lake. Even though our otter might like to join them, males aren't always welcome when a female has young.

The male continues watching this threesome as they move a distance from him and onto shore for breakfast. Once they complete their meal, they again enter the waters where the two young ones engage in a bout of sibling play. They perform the "Ferris wheel" that he and his sister discovered during autumn of their first year of life.

What fun they appear to be having.

As the male otter watches, maybe he thinks about the simpler life during his own youth.

But adult duties call.

He bounds from the scene to perform marking behaviors, then searches for that other female who does not have a family in tow. He wants to become better acquainted with her. When they encountered each other in the past, sometimes their communication was friendly; at other times, it was not. He will just take his chances. After all, he is larger than she.

He wants to make sure she remembers him when mating season is in full swing. Maybe if he encounters her at an activity center, he can initiate the otter dance.

*　*　*

Cooling breezes of autumn spread throughout the land as colorful aspen leaves quake to their music. This is the season when the deer species engage in their ancient rituals of rut. Although there aren't any large grassy meadows in the male otter's home range for elk performances, there are coniferous forests for mule deer and stretches of riparian habitat for moose.

Our otter appears to have not forgotten the lessons he was taught about moose. At this time of year, he tries to avoid the mature bull he sometimes encounters during his ventures. The otter also does not intentionally encounter mule deer bucks; but, if they happen to meet, these two species are at least amiable to each other.

Our male occasionally encounters another animal that mates in autumn, the unique-looking porcupine. Once he inadvertently got too close to one . . . but only once. He quickly learned to stay away from its defensive pointed quills.

This species is at home both in trees and on the ground. The otter sometimes watches their arboreal performances. Now, he watches their terrestrial mating ceremony, but from a safe distance.

The male porcupine approaches the female; once she is receptive to his advances, their ceremony begins. They engage in a comical dance, while singing their sensual songs. When mating begins, the male does not lean over the female; instead, he sits upright while she kindly turns her quilled tail aside. (After all, she doesn't want to stick him!)

Once impregnated, after a seven-month gestation, a cute little spiny pincushion is born.

* * *

I now concentrate my time in the male otter's home range. Does this mean I encounter him every day? No, that is definitely not the case. He might be located in any section of his extensive designated area, as might I. Spending time in this beautiful environment and during my favorite time of year is not a hardship.

Like the otter, I am constantly alerted to other species, particularly rutting moose and fattening black bear. When Old Man Winter enters our world, the bear will seek out a nice cozy den where it can sleep throughout that harsh season. Now, it busily seeks food to fatten up for its hibernation. I really don't feel threatened by bears. I just prefer not to take any chances of an encounter with their kind.

I continue my slip-and-slide behaviors.

One day, I see otter scat on a log that extends out into the waters. So, like any dedicated wildlifer, I advance towards my find. Walking deeper and deeper into the water, I am careful that it stays below the top of my high boots. At least, that is the plan until I slip into a hole and cold waters begin to encase my body.

Choice expressions emerge from my vocal chords.

I move back towards shore.

Upon reaching land, I slide back into the water and down on a murky section of mud. My sopping body is slathered in a sticky, dull-brown garment. Despite being cold, I decide to go back into the waters to wash off my heavy mudpack. Then, I slowly and deliberately walk back towards my vehicle.

How many species of wildlife are laughing while watching this clown-like human?

* * *

Like autumn, the otter drifts into the next season. This will be the first winter in his own home range, so he must meet new challenges. While moving throughout this now-icy, snow-clad world, he locates a small section of open water in a fast moving stream that flows from the smaller lake into the larger one. Here, he can easily enter into the aquatic environment.

When traveling and feeding in the ice-covered waters beyond, he must use those previously learned under-ice breathing procedures. In addition to those methods, the otter discovers he can also obtain oxygen by nudging his nose into clusters of bubbles formed on the underside of an ice layer.

Sliding is very much a part of his mobility. Whether sliding for fun or for locomotion, he appears to enjoy the endeavor. He slides down snow-clad hills, sometimes ending on ice-covered waters. Then, he continues his movements as far as possible, bounding and sliding until he reaches his destination.

He spends time observing other mammals meeting their winter challenges.

One day, he watches a porcupine make a trough through deep snow and then climb a tree where it is able to gnaw at the bark for nourishment. Another time, his attention is drawn to a chickaree as it moves out of its snug little home that is nestled in a conifer, climbs down the tree, and tunnels through the snow to reach its winter food cache. The otter may think how much easier it appears to be for these two species to forage compared to himself.

Is that really the case?

Realistically, most mammalian species have a very difficult time during the long winter season of cold montane climates. Like the porcupine and the chickaree, however, each one evolves ways to meet its challenges.

One day, our otter watches a beaver pop out of a hole in the ice. This may be a rare occurrence as beavers will emerge from the ice in winter only if the temperature is warm enough and the ice thin enough. However, this species has developed very interesting energy conservation tactics to survive Old Man Winter's harshness.

The beavers' physiological and behavioral traits include keeping their bodies well insulated by their thick multi-layered fur, metabolizing their own body fat, eating their winter food caches, huddling together in their well-constructed lodges, and reducing their activities. Mother Nature lends a helping hand by spreading her soft robes of white across their homes to improve the structures' insulating qualities. (Sounds like a cozy situation, doesn't it?)

What about winter survival tactics of our three species of deer?

Moose have a variable strategy, depending on the habitat. They can move through and tolerate snow depths reaching their briskets, as long as they can find forage. In areas that include or are near old-growth forests that serve as a canopy to filter falling snow from the ground, moose may move to even higher elevations to avail themselves of the vegetation the forest affords. Many elk move to lower-elevation wintering ranges if the snow becomes too deep for them to reach forage. Most mule deer do the same. As with all species that don't hibernate or have winter food stashes, deer must be able to find food throughout the winter . . . and not all individuals survive this most difficult time of year.

Since otters evolved to remain active during this season, our male often watches members of his own species. He appears to enjoy the antics of the female with her two young. One day, he watches the threesome sliding down a hill ending at an ice rink. But before trying their Olympic feats on the frozen pond, they work their way back to the top of the hill and down again.

It looks like they are having such fun. Is he thinking about his own youthful activities?

* * *

The cold snowy season progresses throughout the long difficult months. When the calendar finally reflects the emergence of the new season, Old Man Winter is stubborn and does not want to release his powerful control. So, otter and I, along with all the other species, await the outcome of the seasons' struggle.

Some, particularly migrants who aren't equipped to forage and berth in wintry conditions, cannot sustain the lengthened cold season and perish. Others barely eke out an existence until spring finally arrives.

In the meantime, climate notwithstanding, pregnant female otters must seek out their natal dens and prepare them for the new lives they will ultimately bring into them.

* * *

After a female otter's parturition, there is about a 45-day oestrous period for a male to find and court her. Females may not be receptive every year, so our mature male will have to find that out for himself. However, he may not be the only male otter in this area with mating on his mind. Meanderers can be moving through his home range during this most lustful season.

He'd better work quickly to try to find his lovelies!

Female otters may advertise their condition through pheromone (scent) marking at specific locations, or the coming together of the two sexes may just be happenstance. At least our male has become well acquainted with two females in his home range. Knowing each other should help at this time of year.

The male bounds and slides across snow and ice in his quest. One day, he finds one of his love interests. Their pre-mating behaviors soon lead to a blissful and active honeymoon. Copulation occurs both in the open waters and on a nearby section of freshly fallen snow. Their periods of mating are interspersed with periods of rest. After about three energetic days, their honeymoon ends as the male slides away from his bride to search for another.

It will actually be almost twelve months before our male will see the fruits of his labor. Our otter species' reproductive cycle involves a process called "delayed implantation." After successfully mating, the embryo remains dormant for about ten months before beginning its two-month development prior to birth.

* * *

Melting ice and snow tells us Young Man Spring finally emerged victorious over Old Man Winter. The flow of the river intensifies as snows slither into the blue, adding to the force from melting ice. Mother otters know about the spring run-off and potential flooding danger, so they are careful in the selection of their natal dens. They choose those sites which are high enough above, or far enough from, a water source to ensure safety from rising waters.

At least the semi-arboreal species have a safe haven from escalating flows. Those who reside in the coniferous forests can easily find homes removed from such dangers. One species is the pine marten, a member of the same taxonomic family as our otters.

Smaller than the otter, adult martens average just two pounds. They spend time not only in the pines, but also on the ground. They den in hollow trees or in rock crevices. They, too, experience delayed implantation, and bear one to five young with an average of two. Their young are born blind, helpless, and weighing only one ounce. Their eyes open when they are five to six weeks old. Their learning process is much shorter than that of the otters as they leave their mothers when only three to four months of age. (It must be easier to learn to catch rodents and other small species than to catch fish.)

This is the season of emergence. What better entertainment than to observe young animals peeking out from their homes for the very first time. This is the period of innocence. This is the period of excitement.

Listen to them test their vocal cords. Watch them explore their surroundings. Feel their energy.

It is a time of jubilation!

Even the otter's and my spirits spend time engaging in our own period of innocence. We chase each other across the blue while inventing new games to play. We slide and we glide through this unworldly dimension in our own state of jubilation. Eventually, it's time to immerse and disperse before returning to continue our respective teaching.

We return to Earth to enjoy the children of Mother Nature.

* * *

One season moves effortlessly into another. My favorite season, autumn, once again makes its stately appearance. Shades of red and yellow spread their hues across the land as the golden grasses sparkle in the dew.

I sing out to the world: "Ah, this masterful painting, accompanied by Nature's ancient musical score, makes life worth living. It must continue throughout time."

CHAPTER TWELVE

Life's Final Journey

Soon our spirits will remain on high;
through the blue, we will always fly,
playing amongst the fluffy white,
free of danger and of fright.

First, we must end our work on Earth,
thus completing our final worth.
Then, when all these deeds are done,
we will end, immersed as One!

Life's Final Journey

Autumn passes uneventfully, but rather quickly, boding an early and harsh winter. This serves only to corroborate the long-range forecasts. Golden leaves from aspens cloak the surface of Mother Nature's flesh too soon to enjoy them while still abranch.

The two spirits help guide the otter and me through the abbreviated autumn and into the early onset of winter.

It will be a particularly difficult time for all living things. The predicted heavy snowfall will also mean excessive runoff from spring melts, once warm temperatures again emerge. In fact, the entire Earth is in a state of fluctuation.

Is Mother Nature sending a message? Is she striking out at Homo sapiens' devastation of her being, of her children, of her soul? Will those who are causing her grief finally recognize their destructiveness before it is too late for all living species, including their own?

* * *

Our otter seems to be handling this one particular montane climatic extreme quite well, all things considered. The genes passed onto him by his mother and father help him adapt. He probably doesn't like it, but he will call upon his genetics, learned behavior, and species-specific instincts to handle the situation. "Survival of the fittest" is his motto . . . and his own offspring will benefit.

As winter progresses, the beauty becomes much less visually appealing.

So he bounds and slides through his world of ice and snow

What about trudge-along me? Shoveling out from my den every morning and back into it again at night takes a physical toll on my being. (By the end of winter, the accumulated snow pack at my cabin is nearly 4 feet— almost as tall as I am!)

Reluctantly, I decide to move to a lower elevation where I will patiently await spring before returning to my special one's home range to search for the proud father's pups.

I feel confident that his spirit will ensure that happens.

* * *

The austere winter slowly manages to yield to the onset of spring. The male otter survives the rough ordeal.

What about the females from whom his pups are to emerge?

Hopefully, he will at least have one family of which to be proud. Even more would be nice too. After all, his genetic code should be passed through to the next generation and to the generations after that.

* * *

Later spring would make travel to the mountains easier for me, but I decide not to wait. Instead, I return to observe Young Man Spring's defeat of Old Man Winter.

"I hope that conflict won't last too long," I ponder while driving the maintained road surrounded by deep banks of snow. "Where did it all come from? Old Man

Winter must really have been angry this year," I laugh, turning onto the road leading to the entrance of my long driveway.

That is as far as I drive.

"Now, comes the challenge to reach my cabin," I contemplate while strapping snowshoes to my feet and grabbing a shovel for my attempt to get through the door.

Like the beavers' lodge, my cabin also received a helping hand from Mother Nature's insulating blanket of white. When I finally enter my domain, it is warm enough inside for me to endure.

Difficulties aside, I feel happy to return to the Rockies.

Next on my agenda will be reaching my special one's environment.

* * *

April melts into May.

Soon, it will be time for otter pups to emerge from their natal abode.

In the male otter's home range, I discover adult otter tracks leading towards a potential den site. I recheck the location over the next several days and continue to find fresh signs in the surrounding area. Because the otter tracks always lead to the same den site, I conclude it is a female, quite possibly with her young tucked within her cozy home. Oestrous females with nested pups have a very busy time.

Sensitive to the otter family's need for undisturbed privacy, I remain a comfortable distance away.

Does the male otter do the same? Does he know where his pups might be residing?

He certainly senses there is a female in this area since mating season is in full swing.

In addition to the young of various faunal species dawning a new day, another kind of emergence is taking place. As ambient temperatures begin to rise, so do water levels from melting ice and snow. Ponds materialize where there was once dry land. Water seems to be everywhere.

The otter female, on whom I check, chose her natal den at a level that is high enough above the water to preclude it from entering her domain. Only a massive flood can create danger for her family and it doesn't appear that will be the case this year. However, I will keep a watchful eye on the area to ensure that does not happen.

My spirit will let my special one know that he will at least have one family of which to be proud.

Now, we both must patiently await the emergence.

Ah, but what a beautiful season in which to wait with the blossoming of buds into the blooms of life.

Clusters of green peek through brown grasses as small buds dot the bonnets of willows. Tiny flowers begin to awaken from their winter sleep, sprouting bonnie colors. Deciduous trees are shooting out their leaves of green. Accompanied by winds and flowing waters, the avian chorus sings its musical delights. A robin picks up nesting material from emergent flora cradled atop a beaver dam. Snow-capped mountain peaks, which are often shrouded in clouds, now majestically appear above the tops of pine-covered slopes to proclaim their own beauty. Various species of fauna are starting to introduce their offspring for Mother Nature's approval. This is the natural parade of life.

May the beauty of Nature
in all of her glory
be spread throughout time
in a never-ending story.

* * *

The male otter's spirit lets me know that my special one did indeed find an oestrous female. I am also informed that he had more than one energetic honeymoon.

"This means more pups and more good genes entering the otter's world," I sing out in my joyful state. "Yes, it appears the male otter has been making rather merry."

The time for pup emergence in the Rockies arrives. I observe the experiences and explorations of the babes that dawn from the known natal den. I forgot how small they are at this early stage of life. Remaining undetected to their senses, I embrace them with my imagination.

"They look so cute and cuddly," I whisper to myself. "I so wish I could pick them up and hug them."

Realistically, I know I can't do that. They must remain wild and free. They must go through all of the trials and tribulations of their ancestors. They must try to survive.

Much of the behavior I observe between mother and young corroborates that which I documented among my special one and his mother and sister after their emergence from the den eight years ago. This time, however, there is no little spirit to show me what is occurring beyond my senses. I will just have to enjoy that which I see, hear, and feel.

Early one summer morning, after Mother Otter and pups return to their natal den, I rest my body at a protected pond near the lake's outflow stream. I watch the blue-tinged waters flowing with an unrelenting force as they head towards their destination.

"Just think of all the events they will see on their journey. Just think of all the untold mysteries they will carry with them," I think to myself.

I pick up a small stone from the shore. Tiny minerals sparkle in the golden sun.

"Just think of the hundreds of years it took to create this one tiny jewel," I contemplate while returning it to continue its own journey.

It came from a past that I will never know. It will go into a future of which I can only dream.

A kingfisher enters the scene in search of a fish to carry to gaping mouths of a new generation. Other birds

fly into view, making first-rate landings; then, after a drink and a bath, they continue their flights. A hummingbird lands for a brief rest, a quick preen, then a hasty departure. The aptly named flycatcher dives from its perch towards the waters below for some tasty prey, then back again to await another meal to answer its hunger pang. The serenity is broken as two scolding jays demonstrate their impressive repertoire. Their calls are interspersed with the cackling of woodpeckers. Songbirds emerge triumphant, presenting their more appealing music.

Regardless of their particular vocalizations, each species plays a significant role in the web of life.

* * *

One month rolls into another, until once again Old Man Winter sweeps into the Rockies.

I reach the point in my life where I can no longer endure another winter and no longer expend the energy necessary to continue my project.

My special one's spirit will now have to keep me apprised of his adventures. At least, I was able to enjoy observing one of the male otter's known families. I experienced, I felt, and I encased within my soul the beginning of these pups' journey into life.

Before driving too far from my study site, I stop my vehicle and look up into the sky. I watch the clouds becoming shaped into our worldly forms, then play and chase each other across the blue. I watch them immerse and disperse.

"There goes the cloud depicting the otter," I speak as tears fill my being. "But what happened to mine?"

I become consumed within a misty ambience. I become entombed within the fog.

* * *

The male otter persists in his life's journey and appears to have some fun along the way. Like the time a young coyote starts to chase the otter, who then turns the tables and chases the inexperienced pup. He'd better be careful though, since the coyote's parents aren't as gullible.

The otter bounds and slides, runs and swims through the seasons. He continues to sire more otter pups. His genes will be well represented in this small section of North America.

Although the male otter has to endure life's struggle, he appears to be continuously guided by another power: that of the great Otter Spirit.

As the years pass, the special otter watches the emergence of various fauna throughout his home range. Young moose splash in the waters as they crisscross to reach willows lining the banks. Mule deer fawns chase each other during their periods of delight. Newly emerged squirrels, chipmunks, martens, porcupines and so many others enter their worlds full of curiosity and playful behaviors. Cute little beavers bob around on top of the waters during their buoyancy period. Chirps of hungry bird nestlings fill the air during the spring and summer seasons. So many of Mother Nature's children dawn their new day in this spectacular natural world. Each species emerges in the complex masterful form needed to adapt to its particular niche.

The male otter also watches the pups of his own species engage in their childish delights. At least one time, does he briefly join one of his families for a period of joy. They swim together in the waters. They move onto a boulder for the otter dance. The pups chase each other across an island of land, then into the pond for the "Ferris wheel." It is a period of true enjoyment!

The otter's spirit continues to share his real self's adventures with me. It brings me feelings of happiness. Through our spirits and my imagination, I join him during his life's journey. I play and swim with his family. I slide across the ice-covered ponds and down the hills of snow, chasing him during winter's delights. It is very special for me to be kept within his world. Through our spirits, it is very special to the otter to have me join him. Neither of us is ever really removed from the other.

Our special bond is eternal.

* * *

The magic of the spirit world is letting the otter and me know that it will soon be time for another change in our lives.

A larger-than-life Raven lands at the window of my small home in the lower elevations to give me the message. Raven magic will be the guiding force of this change. I always knew he would return one day, just as I knew when he flew from me the day I fell in the snow.

I am ready.

I have contributed as much as possible to Mother Nature. I have spent many years researching her children and sharing my findings with my fellow beings.

Now, I am very tired.

I return to the Colorado Rocky Mountains and to the male otter's home range. I am happy that the spirits choose my favorite season of the year—autumn—for the immersion.

Knowing it will be difficult to find my special one, I call upon the powers to ensure that occurrence. I sit my tired, weary body on a small boulder at the water's edge to wait.

All is silent.

Then, an osprey calls to me from the top of a towering pine. I look up at this magnificent messenger and smile. "I understand," I express to the bird.

There is a ripple in the water and an otter appears!

I sit motionless as he drags his body onto the shore. The unique marking on his head tells me it is indeed my special one. I thank the osprey. This time it doesn't fly away.

I speak softly to the otter. "Please don't be afraid."

Startled, but immobile, are the expression and body language of my special one.

We examine each other's beings. We peer into each other's souls.

Years have taken their toll on both the otter and me. Twelve years have passed since I first saw the curious vibrant pup. Twelve years have passed since he first saw this enthusiastic joyful human.

I slowly extend my hand toward the otter. I gaze into his eyes, which now reflect hundreds of years of suffering for his and so many other species, unfortunately, at the hand of one—Homo sapiens.

"Please try to believe that there really are some human beings who have in the past, do in the present, and will in the future attempt to undo the destructive behavior of the many rapacious interlopers into Mother Nature's natural balance." I speak as a flow of tears spread across our beings and into our souls. "Take my hand in friendship."

We hear the Ancient Voices. The otter reaches his paw out to my hand.

This is the sign.

Each of us knows, in our own way, what is about to happen. The osprey calls out its final message. The otter and I each understand the osprey's message, paw on hand.

My warm smile spreads throughout my being, reaching into that of the otter's. I am feeling, hearing, and sharing the beautiful musical score of Mother Nature's ancient symphony. We listen to the Ancient Voices. We are each ready to connect with the Ancients.

As the magnificent one swoops down towards us, it transforms into a beautiful bald eagle. This is the bird that is given the special power to move freely between Mother Earth and Father Sky. Each of us grasping onto a wing of the Eagle, we are gracefully lifted into our final dimension.

* * *

Look up into the sky and watch the clouds becoming shaped into our worldly forms. Watch us play and chase each other across the blue. Watch, but never stop listening to the Ancient Voices. These voices will guide your Earthly forms.

Never stop listening to the joys and to the sorrows of Mother Nature. Never stop listening to the cries of her children. Everything in Nature holds the spirits. Everything in Nature holds the voices of the Ancient Ones.

If you listen, you will be guided. If you care for and respect her, she will reward you. If you try to destroy her, she will turn against you. You are her guardians and she is yours. Therefore, you must each learn to live in harmony with the Mother of us all. She is trying to live in harmony with you.

Look up into the sky. My spirit and the otter's spirit are being guided by the magic of the Raven. We coalesce into One, as we immerse with the great Otter Spirit!

Listen!
Yes, you can hear the Ancient Voices . . .

Epilogue

During the course of my river otter research, Dorothy Wathen, my Blackfoot friend and Medicine Woman, passed away at her home many miles from mine in Colorado. Not knowing, at first, of her recent death, one night I suddenly became violently ill. Every part of my body shook with such intensity and coldness that no amount of warm covers could stop my shaking. Eventually, calmness did enter my being and I fell asleep. While in this state of sleep, I visualized images of white, dancing gracefully in space and time. I thought this frightful, yet beautiful, dream meant that death would soon overtake me. However, after several days, I recovered from my illness and continued my otter research. It was then that I learned from a mutual friend that Dorothy had passed away prior to my illness. This led me to believe that she was trying to contact me. Our mutual friend thought that I had a true "Vision," which is something very sacred and sought and treasured by Native American people. Throughout my research, I always carried the medicine bag Dorothy bestowed upon me, and I still turn to it for inspirational guidance.

During the months before completing the final draft of the manuscript for this book, visions of white again entered my dreams. The images were abstract at first, gradually taking on more form before finally revealing their otter shape. Each night the image entered my dream it played and danced in joyful glee, just like it does in the story. My little otter spirit helped me through many discouraging rewrites and kept me focused on the purpose for writing this book. After completing the final draft of *The Otter Spirit*, the image appeared one more time and showed me the Rocky Mountains, as if to call me back to that early spring morning where the story began. By the time these pages are read, that journey to the Rockies will have occurred, fond memories recalled, tears shed, and hopefully, the current generation of my otter's legacy sighted performing the "Ferris wheel" in the quiet shimmering waters of the beaver pond in my special one's former home range.

And yes, I still hear the Ancient Voices.

—Judith K. Berg
April, 2005

Otters of the World

SPECIES	HOME RANGE
Cape Clawless Otter *Aonyx capensis*	Africa
(Asian) small-clawed Otter *Aonyx cinerea* (or *Amblonyx cinereus*)	Southeast Asia
Congo Clawless Otter *Aonyx congica*	Africa
Sea Otter *Enhydra lutris*	Northern Pacific Rim (Siberia to California)
North American River Otter *Lontra canadensis* (formerly *Lutra canadensis*)	North America
Marine Otter *Lontra felina*	Pacific Coast of South America
Neotropical Otter *Lontra longicaudis*	Mexico to South America
Southern River Otter *Lontra provacax*	South America
Eurasian Otter *Lutra lutra*	Europe, Asia, Africa
Spotted-necked Otter *Lutra maculicollis*	Africa
Smooth-coated Otter *Lutra perspicillata* (or *Lutragale perspicullata*)	Southeast Asia
Hairy-nosed Otter *Lutra sumatrana*	Southeast Asia
Giant Otter *Pteronura brasiliensis*	South America

Global Otter Organizations

River Otter Alliance	6733 S. Locust Court Englewood, CO 80112 USA http://www.otternet.com/ROA
International Otter Survival Fund	Skye Environmental Centre, Ltd. Broadford, Isle of Skye IV49 9AQ Scotland http://www.otter.org
IUCN Otter Specialist Group (International Union for the Conservation of Nature and Natural Resources)	c/o Aktion Fischotterschutz e.V. OTTER-ZENTRUM 29386 Hankensbüttel Germany http://www.otterspecialistgroup.org
Friends of the Sea Otter	125 Ocean View Blvd, Suite 204 Pacific Grove, CA 93950 USA http://www.seaotters.org

References

Anderson, E. 1989. The phylogeny of mustelids and the systematics of ferrets. In *Conservation biology and the black-footed ferret,* eds. U.S. Seal, E.T. Thorne, M.A. Bogan, and S.H. Anderson, 10-20. New Haven, CT: Yale University Press.

Beckel, A.L. 1990. Foraging success rates of North American River Otters, *Lutra canadensis,* hunting alone and hunting in pairs. *Canadian Field Naturalist* 104:586-588.

Ben-David, M., G.M. Blundell, and J.E. Blake. 2002. Post-release survival of river otters: effects of exposure to crude oil and captivity. *Journal of Wildlife Management* 66:1208-1223.

Berg, J.K. 2002. The diet of *Lutra canadensis* in the upper Colorado River system. In *Otter conservation—an example for a sustainable use of wetland,* eds. R. Dulfer, J. Conroy, J. Nel, and A.C. Gutleb, 8-11. IUCN Otter Specialist Group Bulletin 19A.

———. 1999. *Final report of the river otter research project on the upper Colorado River basin in and adjacent to Rocky Mountain National Park, CO.* National Park Service.

———. 1987. Developmental behavior of three African elephant calves *(Loxodonta africana)* in captivity. *Zool. Garten* 57:171-196.

———. 1983. Vocalizations and associated behaviours of the African Elephant in captivity. *Zeitschrift fur Tierpsychologie* 63:63-79.

Bowyer, R.T., J.W. Testa, and J.B. Fargo. 1995. Habitat selection and home ranges of river otters in a marine environment: Effects of the Exxon Valdez oil spill. *Journal of Mammalogy* 76:1-11.

Dulfer, R., J. Conroy, J. Nel, and A.C. Gutleb, eds. 2002. *Otter conservation—an example for a sustainable use of wetlands:*

Proceedings of the VIIth International Otter Colloquium, March 14-19, 1998. IUCN Otter Specialist Group Bulletin 19A.

Fish, F.E. 1994. Association of propulsive swimming mode with behavior in river otters *(Lutra canadensis). Journal of Mammalogy* 74(4):989-997.

Gittlenan, J.L., ed. 1989. *Carnivore behavior, ecology and evolution.* Ithaca, NY: Cornell University Press.

Hamilton, W.J., Jr., and W.R. Eadie. 1964. Reproduction in the otter, *Lutra canadensis. Journal of Mammalogy* 45:242-252.

Haynes, G. 1991. *Mammoths, mastodons and elephants: Biology, behavior and the fossil record.* UK: Cambridge University Press.

Grinnell, G. (Bird). 1962. *Blackfoot lodge tales: The story of a prairie people.* Lincoln, NE: University of Nebraska Press.

Kruuk, H. 1995. *Wild otters: Predation and populations.* UK: Oxford University Press.

Lariviere, S., and L.R. Walton, 1998. *Lontra canadensis. Mammalian Species* 587:1-8.

Liers, E.E. 1953. *An otter's story.* New York: Viking.

―――. 1951. Notes on the river otter *(Lutra canadensis). Journal of Mammalogy* 32:1-9.

Mack, C.M. 1985. *River otter restoration in Grand County, CO.* M.S. thesis, Colorado State University, Fort Collins.

Melquist, W.E., and A.E. Dronkert. 1987. River otter. In *Wild furbearer management and conservation in North America*, eds. M. Novak, J.A. Baker, M.E. Obbard, and B. Malloch, 626-641. Toronto: Ontario Ministry of Natural Resources.

Melquist, W.E., and M.G. Hornocker. 1983. Ecology of river otters in west central Idaho. *Wildlife Monographs* 83:1-60.

Melquist, W.E., P.J. Polechla, and D. Toweill. 2003. River otter. In *Wild mammals of North America: Biology, management, and conservation*, eds. B. Thompson and J. Chapman, 708-734. Baltimore, MD: Johns Hopkins University Press.

Nitsitapiisinni. 2001. *The story of the Blackfoot people*. Buffalo, NY: Firefly Books.

Novakowski, N.S. 1967. The winter bioenergetics of a beaver population in northern latitudes. *Canadian Journal of Zoology* 45:1107-1118.

Nowak, R.M. 1999. *Walker's mammals of the world: 6th edition*, Vols. I and II. Baltimore: Johns Hopkins University Press.

Patemaude, F. 1984. The ontogeny of behavior of free-living beavers *(Castor canadensis)*. *Zeitschrift fur Tierpsychologie* 66:33-44.

Petersen, D. 1988. *Among the elk*. Flagstaff, AZ: Northland Pub.

Reid, D.G., T.E. Code, A.C.H. Reid, and S.M. Herreo. 1994. Spacing, movements and habitat selection of the river otter in boreal Alberta. *Canadian Journal of Zoology* 72:1314-1324.

Rezendes, P. 1993. *Tracking and the art of seeing: How to read animal tracks and signs*. Charlotte, VT: Camden House Publishing.

Ryden, H. 1989. *Lily pond: Four years with a family of beavers*. New York: William Morrow.

Serfass, T.L. 2003. Otter action plan. *River Otter Journal*, 7 (2). Denver: River Otter Alliance.

————. 1995. Cooperative foraging by North American river otters, *Lutra canadensis*. *Canadian Field-Naturalist* 109 (4):458-459.

Seton, E.T. 1909. The Canada otter. In *Life histories of Northern animals: An account of mammals of Manitoba,* Vol.11, 817-839. New York: Charles Scribner's Sons.

Shannon, J.S. 2002. Behavioral development of otters *(Lutra canadensis)* in a marine coastal habitat. In Dulfer, R., Conroy, J., Nel, J., Gutleb, A.C. (Eds.): *Otter conservation—an example for a sustainable use of wetlands,* eds. R. Dulfer, J. Conroy, J. Nel, and A.C. Gutleb, 312-315. IUCN Otter Specialist Group Bulletin 19A.

Soma, H., ed. 1987. *The biology and management of Capricornis and related mountain antelopes.* New York: Croom Helm.

Toweill, D.E., and J.E. Tabor. 1982. River otter: *Lutra canadensis.* In *Wild mammals of North America: Biology, management and economics,* eds. J.A. Chapman and G.A. Feldhamer, 688-703. Baltimore: Johns Hopkins University Press.

van Zyll de Jong, C.G. 1987. A phylogenetic study of the Lutrinae *(Carnivora; Mustelidae)* using morphological data. *Canadian Journal of Zoology.* 65: 2536-254

Native American Legends

Blackfoot tipi illustration in Chapter 7. Derived from Nitsitapiisinni, 2001, *The Story of the Blackfoot People*, Firefly Books, Buffalo NY.

"Butterflies." Retold from various sources. *American Indian Myths and Legends*, selected and edited by Richard Erdoes and Alfonso Ortiz, 1984, Pantheon Fairy Tale and Folklore Library, Pantheon Books, New York.

"Coyote Takes Water from the Frog People." Told by Barry Lopez in 1977. *American Indian Myths and Legends*, selected and edited by Richard Erdoes and Alfonso Ortiz, 1984, Pantheon Fairy Tale and Folklore Library, Pantheon Books, New York.

"Spring Defeats Winter." Told by Joseph Bruchac. Native American Stories from *Keepers of the Earth* by Michael J. Caduto and Joseph Bruchac, 1991, Fulcrum Publishing, Golden CO.

Index

A

B

C

D

E

F